Power Grid Operations

Written By:

Eli T. Iceman

The front cover photo by Will Waldron of the Times Union, Albany, NY is used with permission. One of the most famous uses of this photo was in the August 25, 2003, edition of Newsweek magazine, following the August 14, 2003 Blackout.

First published by Dog Ear Publishing
4010 W. 86th Street, Ste H
Indianapolis, IN 46268
www.dogearpublishing.net

dog ear PUBLISHING

ISBN: 978-1-4575-1370-1

This book is printed on acid-free paper.

Printed in the United States of America

TABLE OF CONTENTS

Acknowledgements vii

About the Author viii

Preface ix

Introduction 1

Part I:

Parts of the Power Grid & Miscellaneous info about the Grid.

Chapter 1 – Myths and Events 6

Myth 1. - The Electric Grid is aging and deteriorating quicker than we can replace it. 6

Myth 2 – The Power Grid's reliability is extremely vulnerable to an attack. 7

Myth 3. – It takes months to restore the grid if it all goes down 7

Historical Events 8

2011 The Southwest United States Blackout 8

2003 Northeast and Midwest Blackout 9

2003 Italy Blackout 9

1999 Southern Brazil Blackout 10

1977 New York City Blackout 10

1965 Northeast US and South East Canada Blackout 10

Impacts of Power Grid Failures..................11
Aged workforce beginning to retire in enormous numbers....................................13

Chapter 2 – Power Grids = Largest Man-Made Machines Ever Built. 19
Parts of a "GIGANTIC Man-made Machine"................23

Chapter 3 – Generators: 27
Diversity..29
Nuclear 29
Coal...30
Natural Gas...32
Wind Farm..32
Hydro..33

Chapter 4 – The Transmission System................38

Chapter 5 – Substations.......................................48

Chapter 6 – Distribution System.........................57

Chapter 7 – Legislation, Governance and Regulation of the Grids in USA and Canada 67
Public Utility Commission/Public Service Commission...68
DOE – The Department of Energy................................68
FERC – (Federal Energy Regulatory Commission)........68
Homeland Security...68
NERC - The North American Electric Reliability Corporation..69
Regional Reliability Councils..71
RTO's and ISO's – 72

PART II

Real-time Operations

Load Forecasting 79
Tip toe through the Tulips 81
The Last M*A*S*H Episode 81
The Verdict in the OJ Trial 81
9/11 Terrorist Attacks 82
Super Bowls 82
Christmas Lights 83

Inside the Control Room . . . we begin with some building blocks: 87
Frequency 95
The ACE value 97
Constraints and Real-time Contingency Analysis (RTCA) 101
Electric Wholesale Markets 110
SCUC - Security Constrained Unit Commitment........110
SCED - Security Constrained Economic Dispatch 114
Regulating Reserves 120
Contingency Reserves and DCS Events 122

Part III

Information either mentioned earlier or just flat out interesting.

Electric Grid troubles in Japan.......................................126

Smart Grid Technology and what it means to you.......................................130

Emerging Technologies.......................................134

The Environment 140
The What, Where, and How of CO_2 141
Why is CO_2 harmful to earth? 141
What happens if we don't fix it? 142
What is being done to fix it? 142

What does it cost to run some of my appliances? 143
TV 144
Laundry 144
Electric Vehicle 144

Things that may be discussed further in the next Eli T. Iceman book 145
Direct Current (DC) 145
MVAR'S 145
TLR's 146
Demand Response Programs 146

How do I become a Power Grid Operator? 146
Steps involved 147

End Notes 151

ACKNOWLEDGEMENTS:

I have solicited the opinion and advice of many individuals who enjoyed long careers as power grid operators in the electric power grid industry. If there is one thing power grid operators are known for, it is the willingness to share an opinion, whether you want it or not. The individuals I solicited help from did not want published credit for their input, but I wanted to somehow recognize each and every one of them to show how grateful I was for their help in this project. Throughout this book you will notice that I occasionally use the words "we" or "us" when giving thoughts or opinions. I felt I had to recognize their efforts in some way, and this was the best way I could accomplish that while still maintaining their wishes.

ABOUT THE AUTHOR:

Eli T. Iceman is a name that will immediately get the attention of many in the electrical industry worldwide. It is the pen name I have chosen to write this book under in an effort to continue to stay out of the public eye. Power grid operators are at the controls 24 hours a day, 7 days a week, 365 days a year. They are there during an eight hour shift through the middle of the night, or during their kids' Saturday baseball game at the park, or possibly while the family is gathered at the house for a holiday dinner. Nearly every time a person goes to use a light, TV, or anything else that requires electricity, the power is there, in part, because of these people. Thousands of power grid operators throughout the world working in the smallest rural electric cooperative to the biggest regional transmission organization perform their stress-filled job with expertise, professionalism, and an understood lack of recognition by the general population.

Why the name Eli T. Iceman? What does it mean? "Eli the Ice Man" is a memory aid for electrical industry personnel, including engineers, power grid operators, industry leaders, or students of electrical engineering to use for a mathematical relationship between voltage and current. For the person with little to no electrical engineering background, this memory aide can most closely be compared to the grammar school English spelling rule we all remember: "i" before "e" except after "c" or when sounded like "a", as in neighbor and weigh; or the ROY G BIV memory aid to help amateur artists recall the colors of the rainbow spectrum (***R***ed, ***O***range ***Y***ellow, ***G***reen, ***B***lue, ***I***ndigo, ***V***iolet).

Eli T. Iceman recognizes and appreciates all power grid operators, whether it is a generator operator at an individual generating station, a lineman who works from the bucket truck, a substation crew in the confines of the fenced-in substation, the system operator in a localized control room, or the reliability coordinator communicating and coordinating with all members of power grids. For all of these employees throughout the world, current, retired, and deceased, Eli T. Iceman thanks you for the lifelong dedication to "keeping the lights on".

PREFACE

Supposedly, *USA Today* is one of the most successful newspapers ever published for one simple reason. It is intentionally written on a seventh-grade reading level. I have read and enjoy *USA Today*. I have no idea if the easy reading rumor is true, but it sure got me thinking...

Over the years, I have had many family, friends, and neighbors ask: "What do you do at work?" When I was new to the industry I would make attempts to explain it, and I found that I failed miserably each and every time. As time went on, I would hear my colleagues struggle to explain what we do as well. At some point someone started using the comparison of an air traffic controller's job: "He does it for planes, and we do it for electricity." That helped "outsiders" to understand the industry a little better, but any further explanation became blah, blah, blah, and you could begin to see the glassy-eyed bored look on the person's face.

This book is intended to give people a chance to understand the operation of the electric power grid at a non-engineering, basic level. My colleagues and I have done some simple analysis to determine where, if the average Joe reads this book, he would then be in comparison to the knowledge of the general population regarding an electric power grid. After you read this book, we believe, you will know more about electric power grids than a *huge* majority of the general population.

I am confident this book will appeal to all who have just a bit of interest or curiosity in electricity. The interested person could be your neighbor next door who complains about his electric bill, or a new student considering a career opportunity. The book might interest an employee of an organization representing one of the pieces of the industry who would like to know how all of the other pieces fit together. The interested person could be a family member who wants to know just a bit more what his brother-in-law does at work. The book will also be interesting for retirees from the electric power industry, as they will enjoy the book for the memories of the past and will enjoy the possibilities of what might be coming in the near future.

The existing books about the electrical power grid are written by accomplished individuals with enough degrees and awards to decorate their homes from one end to the other. This book differs from those in three main areas. The first is simple: Those books were written by smart, and I mean really academically smart, people. This book was written by me and by others, who have relied less on academics and more on decades of operations experience. The second area is details. Even though some industry books boast of teaching this complicated industry in basic terms, they struggle to do it, and, frankly, after completing this project, I now understand why. It's very difficult to explain the industry's extremely complicated infrastructure in simple terms. In this book, the reader will not be overwhelmed with equations, formulas and technical jargon gained from years of education, training and experience. The final way this book will be different from the other books is that it is written on a more "down to earth," fun reading style. It is a fascinating industry, but it is so technically complicated that a 90-second conversation about it can absolutely bore you to death! The very best efforts were made to add the familiarity of everyday life to the book to help you understand. You might even laugh a few times while reading.

Introduction

EVERY JANUARY, THE SITTING PRESIDENT stands in front of Congress (and the nation) and delivers the State of the Union Address. Without fail, every year, he mentions the power grid, either directly or indirectly. On January 24, 2012, President Obama talked about "repairing America's infrastructure." In that section of the speech, he said we have "a power grid that wastes too much energy." My colleagues and I believe he is partially right and partially wrong, for reasons you will learn by the end of the book.

The underlying intention of this book is to provide a basic understanding of how the electric power grid works and what the power grid operators are doing while they are at work. First, you will walk through the power system from the generators, which create the electricity, through the process of how electricity gets to your home so you can watch TV and drink a cold cocktail.

Before you learn what power grid operators are doing while they are working, you will need a very basic understanding of the power grid's elements. These elements are explained in Part I of the book. Part II of the book is primarily dedicated to the actual responsibilities of power grid operators and the organizations they represent. You will be placed in the power grid operator's chair and learn what he or she would be monitoring and the concerns and corrective actions that he or she would be taking.

While reading this book, you will also learn a few miscellaneous trivia facts you would not normally expect to learn, such as (1) the dire situation Japan's electrical grid is in; (2) the estimated financial costs of large-scale electric blackouts; (3) what the future of smart-grid technology might mean to you in the future; (4) the cost to you to run some of the things in your house, like your TV; and (5) that one of the most important current concerns of industry leaders is the enormous and challenging situation of retirements throughout the industry over the next few years.

During the research for this book, a powerful letter was discovered that was written in 1967 to the President of the United States by an organization named the Federal Power Commission. The letter was the introduction to a 200-page report written in response to the 1965 Northeast US and Southeast Canada blackout that affected approximately 30 million people. I found it interesting that these political leaders nearly 50 years ago had the vision for what the power grid has evolved into today. The content of this letter is included below in its entirety, but remember before you read it, this was written by lawyers and career politicians; that's a polite way of warning you that the letter sometimes gets a little dry.

> *Dear Mr. President:*
>
> *We are pleased to transmit our final report on the Northeast power failure of November 9–10, 1965 and our recommendations for enhancing power system reliability in the Northeast and elsewhere in the Nation. This report is in response to your memorandum of November 9, 1965, and the strong interest evidenced by Congress and the public at large is seeking the causes of the failure and the steps necessary to prevent recurrences. It supplements our December 6, 1965 analysis of the power failure and our interim reports of April and November, 1966.*
>
> *At the time of the cascading failure, questions were raised as to whether a fundamental error had been made in permitting utility systems to be so dependent on other systems that the failure of one could jeopardize great areas of the country and conceivably the entire Nation. The various segments of the industry, responsible observers including scientists, technicians, and academicians and interested governmental agencies, are in agreement that the concept of coordinated efforts is not only a very successful means of achieving lower cost power, but actually enhances the reliability of participating systems. The key lesson of the Northeast failure and*

the subsequent cascading outages, we believe, is that these interconnections and the coordination of diverse systems must be strong in order to be effective.

Since November 9, 1965, many of the more easily achieved improvements in emergency equipment, communications and controls have been initiated or completed in the Northeast and throughout the Nation. These steps will be helpful in avoiding a collapse in power supply. Should failures occur, however, those systems which have improved their equipment will be able to restore power more quickly. Attention has also been focused upon providing emergency power for the essential services during outages of commercial power. But as constructive and as encouraging as these improvements are, we believe the prevention of crippling power failures requires much more.

The Nation's electric power industry consisting of over 3,500 utilities of diverse size and ownership—investor owned companies, municipal and state-sponsored systems, rural electric cooperatives, and federal systems—generally has a superb record of supplying low-cost power on a reliable basis. We are, however, so dependent on continuous electric power in our urbanized and industrialized society that additional efforts must be made to reduce even further the likelihood of potentially hazardous area-wide power failures. Our studies, in which experts from every segment of the industry assisted, convince us that the prevention of major cascading failures of the type which hit the Northeast in 1965, and very recently in eastern Pennsylvania, New Jersey and the Delmarva Peninsula, requires expanded coordination among all utilities on a region-wide basis and a substantial strengthening of transmission systems. We believe implementation of the recommendations in this report—and, as noted above, many are either totally or partially in force today—will substantially assist the industry to achieve these objectives.

As you are aware, the commission, earlier this month, submitted to the Congress a proposed "Electric Power Reliability Act of 1967." That bill would authorize the FPC to play a role in accomplishing a number of the recommendations contained in this report and, in large measure, the bill rests upon the same studies and analyses which underlie this final report on the Northeast power failure of November, 1965.

The Senate Commerce Committee and the House Interstate and Foreign Commerce Committee have demonstrated deep interest in the general problem of reliability of electric service and, accordingly, we are sending copies of our report to the members of those committees. Additionally, because the report contains some information and suggestions applicable to state and local governments, copies are being sent to the governor of each state.

We are grateful for your personal and continuing concern for power reliability, evidenced in so many ways including, of course, references in the State of the Union message and in your Special Message to Congress on Consumer Problems. We are confident that our recommendations will result in more dependable electric service to the Nation's power consumers—a goal that you have articulated, that the Congress has supported, that this Commission and state regulatory bodies must support, that the industry has traditionally given the highest priority, and that the public has a right to expect.[1]

Respectfully,

Lee C. White, Chairman
John A. Carver, Jr., Vice Chairman
Lawrence J. O'Connor, Jr. Commissioner
Charles R. Ross, Commissioner
Carl E. Bagge, Commissioner

PART I.

Parts of the Power Grid and Miscellaneous Information about the Grid.

CHAPTER 1.

Myths and Events

THERE ARE NUMEROUS MYTHS, MISCONCEPTIONS and mysteries about the power grid which circulate through the media and are occasionally in magazines, on radio talk shows or in television news shows. Major events, such as the ones you will read about in the upcoming pages, often cause intense media interest for a brief period of time and revisit some of the questions about the power grid. Some of these myths circulating and discussed are listed below.

Myth 1—The electric grid is aging and deteriorating quicker than we can replace it.

This is simply not true. The grid is aging, but isn't everything? It's updated constantly, and as soon as a component is updated, it begins to age. Common sense should tell you this myth is ridiculous. The world is changing so quickly that electricity has evolved from a convenience to a necessity. Our lifestyles demand a reliable constant supply of electricity. Government and the electric companies combined recognized this long ago and have built safeguards to make sure the system is reliable (you will read more about this in the governance section). What is aging, and will hopefully soon deteriorate, is society's attitude toward the grid, and this attitude is something that desperately needs updating. To be fair, it is actually starting to change, but very slowly. Nobody wants a new power plant or transmission line built near their home in an effort to expand the power grid, including me. Well, there is only one alternative: We need to use less electricity, not more as we have done year after year for decades. Efforts are under way to make the power grid "smarter" and allow the people that use it to have more options of when they want to use based on price. This is one element of an initiative called a Smart Grid and will be elaborated on later in the book. The power grid is never going to be perfect, but take it from some guys that have been around it for a lifetime: Right now, it's pretty darn good in North America. Changes are needed, and changes are coming, but the power grids throughout North America are currently very dependable.

Myth 2—The power grid's reliability is extremely vulnerable to an attack.

Well, the answer to this is how you look at it. Are there parts of the system that are vulnerable to attack? Of course; for Pete's sake, it is ridiculous to think otherwise. We all have seen the attacks on our nation, including the World Trade Center towers on September 11, 2001, so we know these deliberate acts are possible. But as you keep reading, you will learn the grid is an enormous infrastructure. If some parts on it "break," the grid as a whole still works, so the probability of a complete power grid failure because of an attack is extremely low. Also, there are numerous safeguards built in the governance of the grid to help further reduce the already extremely low possibility. So, in other words, possible, yes; probable, no.

Myth 3—It takes months to restore the grid if it all goes down.

While the grid is an extremely complicated infrastructure, it can be started back up reasonably quickly. So what is the definition of reasonably quickly? The grid is run by highly skilled operators in many different locations throughout its area. These highly skilled operators routinely attend coordinated training that simulates different scenarios of how to restore the grid following events—think of astronauts and their training simulators, for example. The scenarios the operators are presented during these simulation exercises are based on careful, thoughtful analysis by extremely intelligent engineers using data from various sources, including previous events. Pin us down and say, "Answer the question! How long?" Well, depending on the event that caused the grid failure, the grid will be back to normal in less than three days and, in most cases, less than that. It is a strong statement, but history supports the answer, as you will see in the next section. The reason/cause for the initial failure plays a *huge* role in restoration. If we are talking about a single electrical event causing a series of failures, it may be rather quick to restore. If we are overcoming some other type of natural disaster or physical attack, well, things could be more difficult to restore.

Historical Events

It is important to understand the definition of electrical reliability. There are really two definitions as reliability pertains to the electric system. One of them refers to the distribution system (which is explained in Chapter 6). Although we will explain the distribution system further, the term “reliability on the power grid” refers to the generators, transmission system, substations and power grid operations all working in unity. We use the term “capital R reliability” as an actual defined term to mean just that: the reliability of the bulk electric system (BES) and its components. Each of these parts will be explained in the chapters ahead, but it is important to understand that if your lights unexpectedly go off, it is most likely because of an unexpected condition in the distribution system, not in the transmission system. When your lights go out because of a problem on the transmission system, then it is quite likely going to get some local media attention or, if it is a large enough event, then possibly some national attention.

Over the past 50-plus years, power grids have been extremely reliable and stable. It is only the few events in those five-plus decades scattered around the entire world which cause people to have a questioning attitude about the grid and to be more curious about it and how it works. Nearly all of the events listed below have created short-term media frenzies and been extremely helpful in the development of the standards, procedures and policies that the electric power grid organizations (and their operators) are responsible to operate to. Below are some examples of major blackout events over the years.

2011 Southwest United States Blackout

Approximately four million customers lost power in parts of Arizona,

Baja California, Mexico, Southern California and the entire San Diego area on September 8. The triggering event was an unexpected loss of a high-voltage transmission line; however, many other factors prior to the line going out of service contributed to the event. There was initial concern from the media that the cause of the event was related to terrorism. The reason for this concern was that the 10-year anniversary of the September 11, 2001 attacks was only days away. The electric companies in the area and the reliability coordinator for this region quickly ruled out the possibility and communicated that conclusion with authorities. It took less than 24 hours to restore all of the customers back to normal. Authorities who investigated the event issued a report on May 4, 2012, stating: "*The September 8, 2011 event showed that the system was not being operated in a secure N-1 state; This failure stemmed primarily from weaknesses in two broad areas—operations planning and real-time situational awareness—which, if done properly, would have allowed system operators to proactively operate the system in a secure N-1 state during normal system conditions and to restore the system to a secure N-1 state as soon as possible, but no longer than 30 minutes. Without adequate planning and situational awareness, entities responsible for operating and overseeing the transmission system could not ensure reliable operations within System Operating Limits (SOLs) or prevent cascading outages in the event of a single contingency; As demonstrated in Appendix C, inadequate situational awareness and planning were also identified as causes of the 2003 blackout that affected an estimated 50 million people in the United States and Canada.*"[2]

2003 Northeast and Midwest Blackout

Approximately 55 million people lost power in parts of Ontario, New York, Michigan, Ohio, Pennsylvania, New Jersey, and New England on August 14. The triggering event was attributed to high system loadings on the transmission system in the area of northern Ohio. Lack of recognition by computerized tools and/or power grid operators of the unfolding deteriorating system conditions caused a series of cascading failures. It took approximately two full days to restore all of the customers back to normal.

2003 Italy Blackout

Approximately 56 million people lost power in Italy on September 28 when a transmission line from Switzerland to Italy unexpectedly was forced out of service. This outage caused overloads on other transmission lines supplying power to Italy. The loading on these lines was more than the lines were designed to handle, so they then went out of service. These cascading failures came just a few weeks after the controversial prime minister of Italy, Silvio Berlusconi, publically broadcast that a disastrous blackout event like that in the United States on August 14 could never happen in Italy! Power grid operators and industry leaders throughout the United States and Canada were silently relieved to have some of the negative media attention shift from the United States to Italy. It was easy for the media to shift attention because of the prime minister's strong comments just days before Italy's event.

1999 Southern Brazil Blackout

Approximately 85 million people lost power in southern Brazil on March 11. The triggering event was determined to be a lightning strike at a transmission substation in the Sao Paulo area. Most high-voltage lines in the station went out of service as a result. This caused a series of generation-to-load imbalances, thus initiating cascading failures. A large percentage of the customers had their power restored within hours.

1977 New York City Blackout

Approximately nine million people lost power in New York City on July 13. The triggering event was determined to be two lightning strikes at transmission substations near New York City. There was some miscommunication among power grid operators accompanied by misoperation of automatic devices and tools the operators depend on during emergency conditions. These factors combined have been attributed to the cause of the event. It took approximately 24 hours to restore all of the customers back to normal.

1965 Northeast United States and Southeast Canada Blackout

Approximately 30 million people lost power in Ontario, New York, New England, and New Jersey. The triggering event was determined to be a transmission line tripped on overload because an automatic device to detect the overload was set incorrectly by an employee months before it caused the event. The device was just sitting, waiting for the "perfect storm" to develop. As a result of this triggering failure, cascading failures occurred throughout the large area, causing the first major power outage of interconnected systems. It took less than 36 hours to restore all of the customers back to normal. This 1965 event initiated the report and letter to the president that you read in the introduction.

To be clear, power grid failures are not the norm. However, to say they are uncommon could be argued. Below is a list of some other major power grid failures.

1967 Mid-Atlantic USA
1978 France
1987 Tokyo, Japan
1997 California
1997 New Zealand
2003 London, England
2003 Denmark and Sweden
2004 Greece
2005 Australia
2005 Moscow, Russia
2006 Large areas in Europe
2006 Tokyo, Japan
2007 Victoria, Australia
2007 Colombia
2008 Brazil
2012 India

Impacts of Power Grid Failures

The general impacts of one of these large-scale events are numerous, up to and including the loss of life. The reasons for the loss of life are primarily due to accidents (e.g., traffic accidents as a result of traffic lights not working), or possibly an elderly or ill person's home life-support apparatus not working. Long-term large-scale outages affect many items that we consider crucial for our everyday lives. Some of these are

—transportation (traffic lights, subways, railroads, airline industry, tunnels, gas stations unable to pump gas, etc.);
—public services (hospitals, water supply, sewer treatment systems); and

—communications (telephone lines, cell phone towers, TV and radio stations).

Now, many, if not all, of these services we have listed are critical to our society, and many have their own backup electrical power supplies as a result of historical lessons learned, but the backup systems only last for short periods of time. The normalcy of everyday life is gone; things are strained and are not operating in their normal manner, so the quicker the normal electrical supply gets back to intended design, the less stressed all of these services will become.

The economic (also called the socioeconomic) impact of these events is an entirely different, but nevertheless very important, subject to evaluate. It is even more difficult to accurately calculate a dollar amount that a blackout event costs society, but it is a dollar amount that industry leaders, politicians and, to a lesser degree, the public are all curious about, so there are people hired to determine the financial impacts of blackouts. Many factors are considered by these expert consulting companies in the final accumulated dollar amount. Some of these factors are

—additional salaries paid to government agencies (for example, overtime to police departments, safety services, and the like);
—emergency health care services;
—lost revenue in the millions of electric meters to electric companies;
—additional salary expenses of the electric industry workforce to bring the grid back;
—loss of refrigeration, causing food to spoil in homes and businesses;
—lost income to workers throughout the affected area;
—lost revenue from factories losing production time of whatever products they make;
—other cost factors, such as riots and looting;
—insurance rate premiums increasing as a result of accidents or other scenarios; and
—many others I am not thinking of.

In the 2003 Northeast and Midwest blackout, these accumulated items were estimated to have totaled between $4.5 billion and $8.2 billion, according to one study done by the Anderson Economic Group

(AEG).[3] Another consulting company, called ICF Consulting, calculated the value to be between $6.8 billion and $10.3 billion.[4] The Department of Energy published its estimates at approximately $6 billion.[5] As you can see, the numbers vary in the studies from a low of $4.5 billion to a high of $10.3 billion. It is probably impossible to put an exact number on the event, but no matter what the exact value is, we can all certainly agree on one thing: Several billions of dollars for a power grid failure that lasts 24 hours is an enormous incentive to learn from mistakes and take corrective action.

Aged Workforce Beginning to Retire in Enormous Numbers

It is an industry fact that over the next 10 years, the electric industry's workforce is expected to be solidly impacted by retirements. Industry leaders and politicians alike are aware of the situation and have growing concern about it.

The federal government's Bureau of Labor Statistics states on its website some significant points about the electric industry's workers:

> —Job prospects are expected to be excellent for qualified applicants as many workers retire;
> —Earnings are higher in this occupation than many other occupations that do not require postsecondary education;
> —Familiarity with computers and a basic understanding of science and math are helpful for those entering the field;
> —Several years of on-the-job and classroom training are required to become fully qualified; and
> —A growing number of industry retirements should create very good job opportunities, especially for electric power grid operators, power-line installers, and repairers.

The website also states:

> *Power distributors and dispatchers, also called load dispatchers or systems operators* [these two positions we refer to as power grid operators throughout the book] *work for utility companies, non-utility generators, and other companies that access the power grid. They control the flow of electricity through transmission lines to industrial plants and substations that supply residential and commercial needs for electricity. They monitor and operate current*

converters, voltage transformers and circuit breakers. Dispatchers also monitor other distribution equipment and record readings at a map board—a diagram of the transmission grid system showing the status of transmission circuits and connections with substations and industrial plants. In doing this, they communicate closely with power plant operators, energy traders and local utilities to route energy from generating stations to customers.

Job prospects. *Job opportunities are expected to be excellent for well-qualified applicants because of a large number of retirements in the electric power industry. During the 1990s, the emphasis on cost cutting among utilities led to hiring freezes and the laying off of younger workers. The result is that many power plant operators, distributors and dispatchers are nearing retirement age. Utilities have responded by setting up new education programs at community colleges and high schools throughout the country. While many individuals are showing interest in these high-paying jobs, prospects will be best for workers with strong technical and mechanical skills, and an understanding of science and mathematics.*[6]

Some studies show approximately half of the industry workforce will be at retirement age over the next decade. This is a concern to many politicians and industry experts throughout the industry, because losing the experienced workforce is an ugly recipe for trouble. Here is a quote discovered from a long-term reliability assessment on the North American Electric Reliability Corporation (NERC) website: "The loss of industry workers and their years of accumulated expertise due to retirements is a serious threat to the bulk power system reliability *exacerbated by the lack of new recruits* entering the field."[7]

Let's elaborate on this thought.

The power industry has been under pressure to improve reliability for more than half a century. Much of this pressure was basically political and peer pressure and did not have any real financial repercussions; however, as politicians, industry officials and the public reacted to the 2003 blackout, this situation changed. As a reaction to the human and mechanical errors that contributed to the 2003 blackout, the federal government gave authority to an organization called North American Electric Reliability Corporation (NERC) to impose fines for violations to the standards or rules that electric companies and grid operators are required to operate the power grid under. These financial penalties can be *extremely* significant, and some companies have recently incurred penalties that exceeded tens of millions of dollars.

With many retirements expected throughout the industry, the experience base in generating stations, substations, and control rooms will be greatly reduced. As the situation progresses, there will come a time where less experienced operators remain to pass on the knowledge needed during on-the-job training of new recruits. The new recruits end up replacing their mentors and begin teaching the people that replace them. A lifelong utility employee, Ralph Paride, in North Carolina, predicted this scenario more than a decade ago and said it is a troubling situation that is out on the horizon. Well, he was correct. He called it the rule of 80s because generally you cannot teach someone every single thing you have learned over a career, but you can comfortably teach a new recruit 80% of what you know. If there is not enough time for the new recruit to experience the other 20% before they begin to start passing on the knowledge, they will be passing on 80% of what they learned. Now remember, they only learned 80% of what the original "old-timer" knew, so in effect, the new, new guy just got 64% of the old-timer's knowledge. Here is a typical example of what it looks like all summed up, using a three-man team working in an electric power control room, generating station, or part of a three-man line crew traveling in a bucket truck:

> Old-Timer (39 years' experience) has 100% skills and knowledge for the job.
> New Recruit A (6 years' experience) has learned 80% of the skills and knowledge for the job.
> New Recruit B (2 years' experience) is learning the job.
>
> Old-Timer retires and enjoys hunting, golf, traveling and the grandchildren.
>
> New Recruit A gets promoted, New Recruit B gets promoted, and New Recruit C is hired.
>
> New Recruit A has 80% knowledge base of the skills needed.
> New Recruit B gets 80% of 80%. He now has approximately 64% of the knowledge Old-Timer had when he left.
> New Recruit C is learning.

The knowledge base of the team in this particular example is greatly reduced from what it was prior to the old-timer leaving. Even though experienced personnel make mistakes, a mistake is obviously more likely to occur from inexperienced personnel. You can see from above (even if you argue the percentages) that the knowledge and skill base are diminished, causing mistakes to be more likely. Now consider the tens of thousands of anticipated industry retirements throughout North America and you can begin to understand the concern of the industry's leaders. Have you ever heard the saying "You learn from your mistakes"? The old-timer might have already seen the mistake made before and be able to stop it from being repeated, and if New Recruit A hadn't ever encountered the situation, he might make the mistake the old-timer could have prevented. The old-timer provides on-the-job training (OJT) that includes an invaluable base of personal experience obtained from years at the position. He can share the knowledge of his past mistakes or triumphs of the things he did correctly. Many intangible things are also passed on during OJT, such as how to look for the unexpected or the abnormal conditions and how to have a questioning attitude toward these occurrences. The only way other than experience to reduce the probability of human errors is with classroom training, which can be effective but is not usually to the level of OJT with an experienced operator.

The industry has been improving its operational training programs for years and continues to do so. There are three primary reasons for the improved operator training programs over the past several decades. The first is safety. Operators and field personnel must be aware of the dangers to themselves and others from mishaps in routine and emergency maintenance of elements of the grid. The second is economic. It is just too costly for companies in today's economy to pay for errors caused by lack of training or by human error and that could have been prevented. The third is preparedness. Power grid operators must know what to do when certain events or situations occur. Most of these events do not occur frequently, so it is important to review, understand and be prepared to react to such instances. Below is another reference from a PowerPoint presentation on the NERC website. It really summarizes the industry's concerns about training and the serious retirement issue that is looming. It is one of the proactive approaches to address the known problem.

> *NERC is also coordinating the efforts of various industry participants, the Idaho National Lab and the Pacific Northwest National Lab in developing the North American Grid Center of Excellence, which will be an enhancement to existing operator/dispatcher simulators. This enhancement will improve the ability to transfer knowledge from the aging workforce to the incoming workforce. The enhanced simulators will also be valuable in validating planning models, event reconstruction and analysis, and is consistent with recommendations in the final report of the U.S.–Canada Power System Outage Task Force on the August 14, 2003 Blackout.*[8]

So, how did the electric industry get into this retirement dilemma to begin with? It is a combination of factors over the course of a few decades that include hiring freezes throughout the industry to reduce budgets and increase shareholder profits in the '90s, as well as company mergers in 1998–2000 and again in 2004, which caused duplication of efforts in some job functions. Therefore, the elimination of less-experienced employees resulted. Also, because of the technology explosion of the '80s and '90s, industry educational programs in colleges lost the momentum they had in previous years. Students with engineering interests saw unlimited opportunities in the technology world of gadgets and leaned in that direction to explore their creativity.

The electrical industry has recently been working with colleges and universities to create educational programs for students to develop the skills required to plan, build and operate the electric grid. An internet search of schools with programs of this nature should provide you with plenty of information if you have further interest. Some schools known throughout the industry that are working with electrical industry leaders are Drexel University, University of Cincinnati, Mississippi State, Pennsylvania State University, University of Delaware, Bismarck State College, and Thomas Edison State College and many others. Even NYU (which is famous for many curricula, including acting) now has a polytechnic institute with electrical power grid operation and electric market courses—actors and actresses on one campus, power grid operators and engineers on another.

Engineering degrees are certainly valuable to have, and you are encouraged to get one if possible; however, I will tell you many of the individuals I solicited for help on this project do not have an electrical engineering degree or even a college degree. All of the individuals have enjoyed long careers in the industry, and all of them are quite intelligent. As you read earlier, the Bureau of Labor statistics states that post-secondary education degrees are not necessary to obtain a high-paying job in the electrical industry. But rest assured this educational requirement will change with time, so for those individuals, who have interest, take advantage of it while you still can.

Later in this book, an outline is provided of one of the many ways to help yourself or someone you know be placed in a favorable position to get a job in the industry for minimal cost compared to a college education. For decades, the power industry had a well-deserved reputation for a hiring practice called nepotism, which means people who worked in the industry would make efforts to get their family members jobs in the industry when they reached adulthood. Think about it . . . why not? It makes sense. Parents felt their jobs had been great for them for years, so why couldn't it be great for their children? I and many of my retired peers can speak honestly about this, as we made attempts to use this well-deserved industry stereotype to get our children jobs decades ago. This nepotism practice has changed throughout the years as the command and authority of legal and human resources departments has grown throughout all businesses nationwide. In our experience, looking back through our careers, we believe the electric industry is an industry that, once you start in, you don't typically leave; it offers competitive salary, stable job security (do you envision living without electricity?), good benefits, and a constantly changing environment to work in.

CHAPTER 2

Power Grids = Largest Man-Made Machines Ever Built

IF YOU THINK OF A power grid as a machine, the understanding of it get easier. In the continental United States and Canada combined, there are only four "machines" that get electricity where it is needed. The machines have names: The two largest are called the Eastern Interconnection and the Western Interconnection, and the two smallest are called the Quebec Interconnection and the Texas Interconnection. These four machines are all separate and do not operate in a connected, synchronized manner.

That's right! FOUR! Think about that. It really is kind of incredible that four power grids provide enough power for approximately 300 million Americans and 30 million Canadians.

Now imagine the map with the two small interconnections removed. We have the Western Interconnection and the Eastern Interconnection. Why are there two interconnections for almost all of the United States and Canada? Why not just make it into one interconnection for the entire country? If you look at the picture and think about it, you can answer the question on your own. We'll give you one hint, though: An interconnection needs to be pretty tightly connected across large areas with lots of transmission lines connecting the major load areas. So, why divide the country into these two interconnections? What is in between the Eastern Interconnection and the Western Interconnection if you

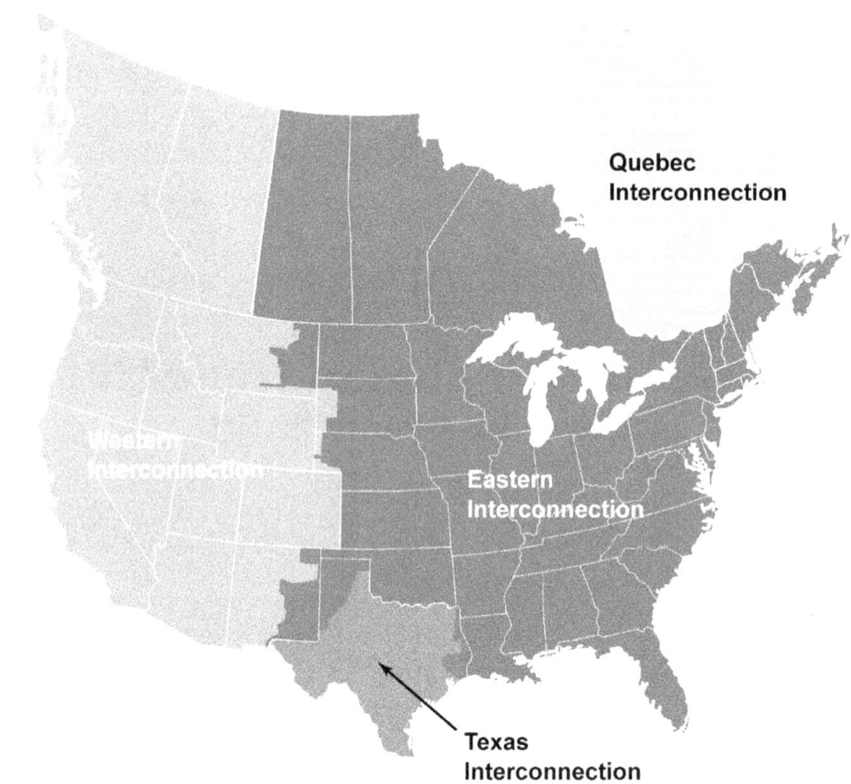

Four Major Interconnections in the Continental US and Canada   Illustration I-1

and your family were driving from one to the other? Hopefully, you came up with the answer, but it's pretty easy to get this one wrong. If you guessed the Rocky Mountains, you were only partially correct. The Rocky Mountains run down through the Western Interconnection (WI), but the major cities in Colorado, New Mexico and Montana are actually tied into the WI. What actually has a bigger effect on the separation of the two interconnections is wide open spaces. The grid was basically built from the coasts inward, and when the electric companies tried to tie it all together many decades ago, it was just too weak across the western edges of the Great Plains. At one time, the grid was actually operated with the east and west tied together, but the flows across the limited transmission lines tying the systems together would cause more operational trouble than they were worth, so the decision was made separate and operate the two interconnections independently.

An obvious question in your head right now might be "Why is Texas a separate interconnection?" Well, if you have ever been to Texas or met a Texan, you may have noticed a bit of an attitude of "everything is bigger in Texas" or "everything is better in Texas." It's so bad that when a Texan opens a package of M&M's candy, they truly believe the individual chocolate candies are bigger when you open them in Texas than anywhere else. So, Texans basically said, "Our system is bigger and better than the other interconnections." Okay, okay, sorry to all of you Texans—just having some fun at your expense. Truth is, the Texas Interconnection is separate because the state government made a conscious decision decades ago to avoid dealing with interstate commerce regulations of moving electricity across state lines. The organization responsible for the Texas power grid, called the Electric Reliability Council of Texas (ERCOT), does a very good job with the operation of the power grid. With that said, there have been opinions that Texas would benefit someday from joining a larger interconnection. Texans' average kilowatt prices are slightly higher than those in their bordering states. They have recently created and formed their own market, but currently, there is not enough public historical data to evaluate whether it is successful yet.

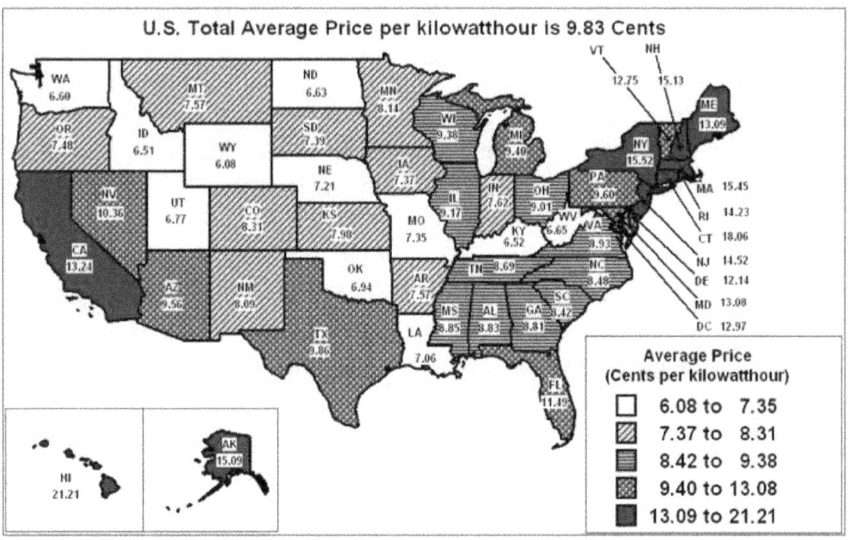

Note: Data are displayed as 5 groups of 10 States and the District of Columbia.

Source: U.S. Energy Information Administration, Form EIA-861, "Annual Electric Power Industry Report."

Source: U.S. Energy Information Administration Illustration I-2

Why is the Quebec Interconnection separate? Many of us have gotten the opportunity to train with Quebec operators and visit some of their generation plants and substations throughout the years, which helped me to know how to answer this question. Basically, the reason that they're a separate interconnection is because geographically, Quebec has very few areas with a population density near the Eastern Interconnection. Because of this, there are very limited areas to tie the two interconnections with transmission lines together. This makes for a weak interconnection. Another reason is they have a lot of hydro generation that is located very remotely (hundreds of miles) from where their load is located, requiring special electrical needs to run the system reliably under certain conditions. Quebec's system is set up slightly differently than other interconnections because the conditions described above often create electrical anomalies on the grid: which the Eastern Interconnection isn't set up to operate within under normal conditions. However, the Quebec Interconnection is tied into the Eastern Interconnection with a different type of technology of transmission, which is called direct current (DC) ties, to make these electrical anomalies transparent to the rest of the Eastern Interconnection. The DC circuits are not explained in great detail in this book, as you will be given enough to think about without adding this technology.

So, although the Quebec Interconnection is in reality tied to the Eastern Interconnection by DC circuits, the Quebec Interconnection is considered a separate interconnection, because the only time both interconnections are operating to the same *frequency* is completely by coincidence. Frequency will be explained in greater detail in Part II of the book, so don't worry if you're not sure what the last few sentences mean.

But there you have it: four. Four machines provide electricity to all of the homes and businesses, streetlights, and anything else that uses electricity in Canada and the continental United States.

Since the objective is to explain to the reader about basic power grid operations, we take the picture and remove three of the machines. Here is a picture of just one of the machines, the Eastern Interconnection. This will be the machine discussed further. The other machines operate in a very similar manner, so you can apply most of what you learn about this one machine to the others. There may be some subtle differences, but for the big picture, the interconnections really are all quite similar.

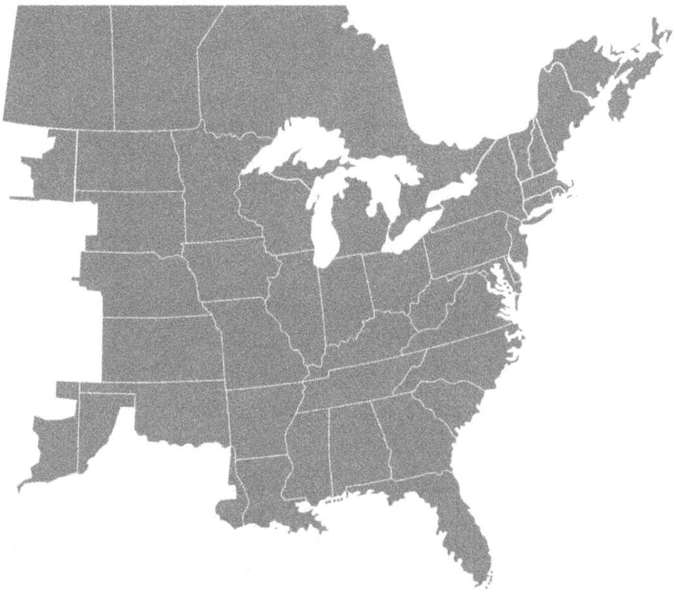

The Eastern Interconnection Illustration I-3

Parts of a "GIGANTIC Man-Made Machine"

Okay, four enormous machines have been identified. In any machine, there are parts. Consider that a bicycle has five parts: a frame, a chain, tires, handlebars and a seat. Can you argue there are more parts of a bicycle? Absolutely. Sprockets, spokes, nuts, bolts, etc. But to make the power grid understandable to the average person, we simplify the machine; therefore, we will divide the parts of an electric grid.

Each of these machines arguably has six parts: generators, transmission lines, substations, distribution system, legislation/governance, and real-time operations.

Before we examine the parts of the machine, it is important to tell you that as you read about several of these parts (the first five), the explanations of them are extremely simple and somewhat incomplete in order to keep it understandable. You can read dozens of books on each of these parts and still not know all there is to know about them, but to understand the operation of the electric grid as a whole, you will have to know what they are, not necessarily exactly how they work. I do not know how to run a nuclear power generator, nor can I climb a utility pole and get your lights back on when something causes them to go out, but I, along with my colleagues, know how all of the pieces fit together, and that is what we will attempt to pass on to you.

Some of the parts I describe will open me up for criticism from industry experts, knowledgeable engineers, experienced operators, and some other industry colleagues. The reason it will do this is because some of these subjects are intentionally oversimplified. But frankly, I am too old and cranky to care about the criticism, as this book is not written for them. What is provided in Part I of the book is the foundation you need to have a general understanding of Part II of the book, which is entitled "Real-time Operations." Part II will be the area where you go a little further than general knowledge and into what many people are curious about.

To summarize, the above:

The five straightforward explanations will be about:
Generators
Transmission System
Substations
Distribution System
Legislation/Governance of the Power Grid

The more in-depth explanation will be about:
Real-Time Operation of the Machine

Below is a very simple illustration of what the power grid looks like.

Illustration I-4 shows generators injecting electricity into multiple generator substations. Those substations provide paths to other substations throughout the grid using transmission lines. The substations are all connected by more than one line for reliability redundancy. The generators feed electricity to the substations. The transmission

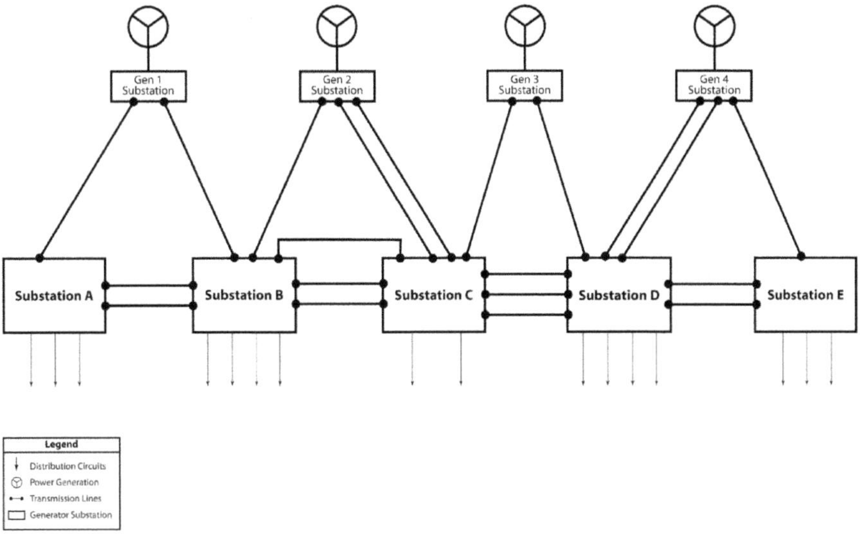

Power Grid Diagram Illustration I-4

lines allow the power from the generators to go to the substations, where it is needed; this is very similar to a city water system, in which the water is in the pipe and goes where it is needed. The distribution circuits coming out of the substations each feed hundreds or thousands of customers. As you can see from the illustration, the distribution lines are not tied together for redundancy as the transmission grid is.

Something different in this illustration that is not accurately represented in other books is a significant teaching moment I feel is worth mentioning. On the transmission system, there is normally more than one source into transmission substations. Electricity sources on the transmission grid are normally not radial, as many other teaching aids insinuate from their diagrams. Other teaching illustrations show a generator with one transmission line to a substation, with a distribution line to a house. These widely used diagrams make it appear that if something goes wrong with the generator or transmission line, the electricity will not get to the house. The electric transmission system, as part of an electrical power grid, is not radial; however, the distribution system is, for the most part, radial. This simply means that once you get to the distribution system, there is only one feed to your house. The two systems are very different, as you will learn in the upcoming chapters. This point may seem somewhat trivial, but to me, if you see a picture diagram and it is not completely accurate, the more you read, the more you may become confused. So, to be clear, the difference between the transmission system and the distribution system is significant, and is important to understand as you learn more.

CHAPTER 3

Generators

THERE ARE MANY WAYS TO explain generators. Technical jargon referring to generators is abundant. It could be used here to easily get you to stop reading at the point you are overwhelmed, which could be in as little as three sentences. For technical purposes, which you don't have to get too focused on, an explanation of how the most common generators (fossil fuel generators) work is given in the briefest of terms. Don't give up on this book because of the next few sentences, as it is not critical to remember the information in order to understand what power grid operators do in their daily job functions. Well, the generator operators throughout the power grid might have good reason to disagree with that comment, but for now, just have some trust.

The large majority of generators on bulk electric power grids are driven by steam. The steam is created by burning coal, natural gas, or oil, or by a controlled nuclear reaction. The next phase is to send the steam at extremely high pressures to a set of blades similar to fan blades or, more specifically, similar to jet engine blades. The steam is forced through the blades to turn a shaft, which has a rotor attached to the end of it. As the shaft turns, the rotor (also known as an electromagnet) turns with it. This produces a rotating magnetic field. As this field moves through coils of copper wire in the generator stator, it is producing electricity for the grid to use.

You need not worry, because that paragraph is as technical as you will read in this book, and if you didn't understand it, no worries; just keep reading. The details of how a generator actually works are not critically important to understand in order to grasp the overall operations of the power grid. However, what *is* critical to remember about generators is in the next two paragraphs, and it is much simpler.

Generators all have a range they can operate in, called an operating range. What does that mean? The generators on the power grid all produce electricity, which is measured in a unit call megawatts (MW). An explanation of a MW will be provided in one of the upcoming pages, but for now just know it is a unit of measurement. Here is an analogy of operating range for generators: The open-wheel Indy race cars have an operating range of 0 mph to 220 mph. Generators on the grid have many different operating ranges based on what design they are, but for example purposes, let's assume we are talking about a 300 MW coal-fired generating power plant. Three hundred MW is the most power the plant can produce. When the plant is offline, it is obviously producing 0 MW, so the range of the unit is 0–300 MW. Well, almost . . . if the unit is scheduled to run, many things happen for 12–24 hours to prepare the unit for operation, but once it comes onto the grid, it immediately starts to inject megawatts into the electric power grid. The unit is designed to have a minimum output level to keep the unit stable. For this example, let's assume the minimum stable output level is 70 MW. That means the operating range of this unit is 70–300 MW. Power grid operators can send instructions to that individual generating unit to operate somewhere in that range.

That was the first thing you need to know about generators, and here is the second: Indy car drivers can put the pedal to the metal and increase the speed of the car from 70 mph to 200 mph in just a few seconds. The particular generator in this example can put the "pedal to the metal" and increase the output of the unit from 100 MW to 105 MW over the course of five minutes. This is what power grid operators refer to as ramp rate or, using the Indy car analogy, acceleration. The ramp rate for this unit would be known to power grid operators as 1 MW per minute. It is important for power grid operators to know the range of the units in their area and the ramp rates of each of those units. Power grid operators—or, more specifically, the software they use—send signals out to generators in their area to increase or decrease generation based on the operating characteristics of each unit as well as the power system conditions at the time. You will learn a great deal more about the changing system conditions that require power grid operators to move generator output up and down in Part II of this book.

Diversity

Generators use all different types of resources to operate. It is important for companies that have a large collection of generation to use an assortment of sources of fuel to maintain the reliability of the grid; this practice is called fuel diversity. If all the eggs are in one basket, things could go wrong quickly (as the stock market has unfortunately taught many of us).

For example, if a company relies on coal for 95% of the fuel it uses to produce electricity and it can't get coal delivered because of a railroad workers' or coal miners' strike, then the company could be in jeopardy of having to shut 95% of its plants down because it can't obtain the necessary coal to use for producing electricity. Or let's say a company decides that wind generation is the way to go and it makes the decision that 65% of its needs will be supplied with wind generation. They are dependent on the wind blowing, but if it is a calm day, there will be no wind generation produced.

By having an assortment of different power plant designs using different fuel sources, risk and opportunity is spread out. Some engineers a long time ago came up with a great idea to have some power plants strategically designed to have the option of using one of two different sources of fuel, such as oil or natural gas. This way if the price for natural gas is considerably lower than for oil, the plant can produce electricity by burning natural gas, but if oil prices are lower than those for natural gas, they can make some slight equipment changes and begin burning oil. Having this option gives companies the ability to choose the source of fuel they will use to maximize profits.

Diversity of power plant designs throughout any large interconnection is one of the most important factors in having a cost-effective and reliable power grid that consumers can depend on day after day.

Below are some pictures and brief descriptions of different electrical-generation plants.

Nuclear Power Plants—These plants make use of controlled nuclear reactions as the source of heat to create the steam used to turn the magnet inside the stationary coils of copper wire. Nuclear power plants have to deal with one additional regulatory organization that other power plants do not: the Nuclear Regulatory Commission (NRC). The NRC gives the approval or licensing to the power plant to operate. After

Nuclear Generating Station Source: Shutterstock.com

license approval, the operation of the nuclear power plant must strictly adhere to the rules and regulations of the NRC. The water used to cool the generator and nuclear reactor is in a closed-loop system that then circulates cooled water in a separate system to cool the water in the first closed loop. The water from the second closed system then goes through radiator-type systems located in these cooling towers that have outside water trickling down over the radiators and cooling the second looped system's water. The water vapor that is billowing from these towers is therefore twice removed from the water circulating in the reactors.

Coal-Fired Power Plants—These plants burn coal in one of two ways, based on the original design of the plant. One old-style design is to have a slow-moving metal conveyor belt with crushed pieces of coal on it to burn at the bottom of the boiler. The more modern way of burning coal is to send the coal through a series of crushers and pulverizers to reduce the coal into a powder form so by the time it is ready to burn, the coal has the consistency of baby powder. It is then mixed with air and blown into a furnace, similar to spraying gas. The coal burns to produce the heat needed in the furnace to make steam used for the mechanical force to turn the magnet inside the stationary coils of copper wire. The coal comes to the plant via railroad, river barges, and/or tractor-trailer trucks.

Coal Generating Plant Source: Shutterstock.com

Natural Gas Plant Source: Shutterstock.com

Natural Gas Power Plants—These plants burn natural gas, which comes to the plant by way of an underground high-pressure natural gas-transmission system. The gas is sprayed into the furnace in the same manner as the pulverized coal. The operation of oil power plants is very similar in theory to natural gas plants as the oil is sprayed into the furnace as well. But the oil does not come from an underground delivery system; it comes from large steel containment tanks that are replenished as needed. As the natural gas or oil is sprayed and burns in the furnace, it, like the fuels in other types of plants, turns water to steam, and the steam creates the force to create the motion to produce electricity.

Wind Generation Source: Shutterstock.com

Wind Farm Generating Sites—Wind-generation farms could be the quickest growing technology of electrical grids today. Selecting sites where these wind farms will be built is not without its challenges. The wind flow must not be restricted, which normally means the wind farms are built in high-elevation areas. They must also be somewhere near a part of the power grid they can attach to so the generation they create can be injected into the grid. The next obstacle is that these sites must be in a terrain where access roads can be built so the heavy equipment for construction can have access. The obvious great benefit to wind is there is no fuel cost. The downside is it is 100% dependent on wind to be operable. Forecasting how much wind generation operators can expect used to be a major operational and economic concern, and it still is, to a smaller degree. The forecasting of how much wind generation you can depend on for an hourly or daily forecast has improved incredibly since 2009. There are wind-forecasting companies that provide this information, and the quality of the forecasted information is continuously improving with time.

Hydroelectric Power Plant Source: Shutterstock.com

Hydroelectric Power Plant—Water moving from upstream to downstream by way of gravity spins the blades in these plants, causing the movement needed to turn the generator shaft and the magnet inside the stationary coils of copper wire. Hydro power is the resource that power grid operators enjoy the most because the level of electrical output can be adjusted very quickly (its ramp rate, as discussed earlier) to respond to changing conditions on the grid. A coal unit might have a ramp rate of 1 MW per minute, whereas a hydro unit is likely to have ramp rate of 25 MW or more per minute.

The preceding images were just to give you a sense of familiarity, as you have probably driven by many generators which look similar to the images. There are thousands of generators scattered across the "machine". There are approximately 7,000 generators larger than 20 MW spread across the Eastern Interconnection. If you consider that each dot on the map below is a generator of at least 20 MW in size, this is what it looks like with the estimated 7,000 units scattered throughout the Eastern Interconnection.

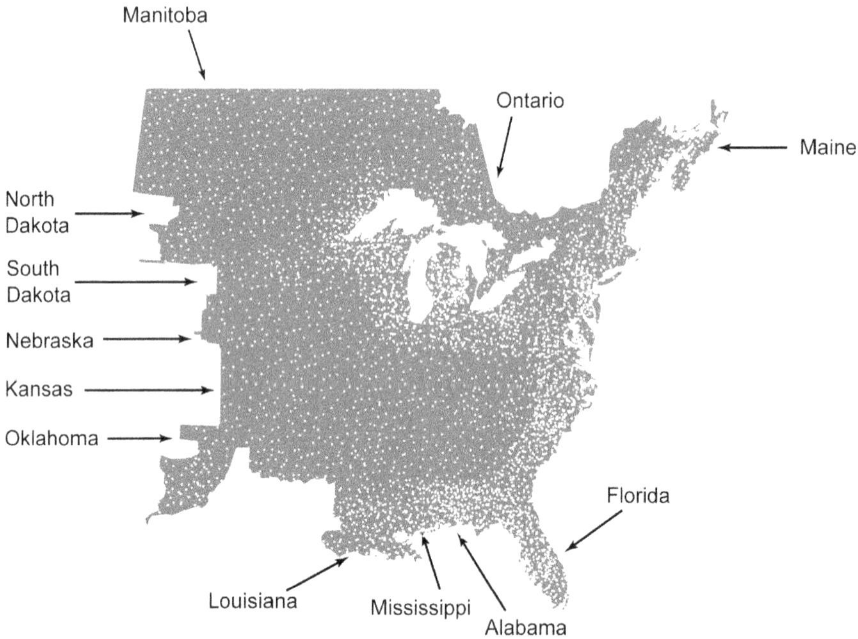

Illustration1-5

Well, actually, there are probably more like 2,000 dots on this diagram, because when I had it created, I thought it looked ridiculous with 7,000 dots on it. It reminded me of one of the morning TV news shows where they give the airline report of flights and show a map where there are planes overlapping everywhere. If you're a frequent airline traveler, that map appears extremely scary until you realize that each plane on the map is larger than the state of Rhode Island. So, I wanted to just illustrate on this map that there are *a lot* of generators scattered throughout the interconnection.

The map is not perfectly to scale and does not accurately represent the geographical locations of all of these 7,000 generators. It is meant to show you how remarkable it is to have that many generators scattered throughout the machine. Triple the amount of dots in your mind to get a better of idea of how many there are. There are also many more generators smaller than 20 MW throughout the interconnection. In all, there are more than 600,000 MW of generating *capacity* in the Eastern

Interconnection.

Generating capacity—what's that? All capacity means is that you add all of the maximum outputs of the units in the interconnection together. Now, many of you who are reading this may be thinking to yourselves, *What is a MW?* Let me explain it in a very untraditional manner.

Generators produce electricity, which is measured in a unit called watts. A kilowatt is 1,000 watts. Kilowatts are the unit of measurement to calculate how much electricity you used over the course of a billing cycle. So here is some simple math for you to figure out how much it costs you, the end user: A kilowatt or any other watt is measured instantaneously but billed hourly—just like speed is measured in miles per hour.

Kids and grandkids often leave 100-watt light bulbs turned on in their rooms. If they leave it on for an hour while they are outside playing with friends, how much money do they waste? Electric utility rates are different across the country, but if we use 10 cents per kilowatt hour, which is a reasonable average value to use across the country, the calculations are just a bit easier. A kilowatt is 1,000 watts, so a 100-watt light bulb uses 1/10 of a kilowatt each second it is turned on. So, if it is on for an hour, the bulb will end up using 100 watts, or 1/10 of a kilowatt hour, or 0.1 kWhr. Therefore, the kid just wasted a penny. So, what the heck are you yelling at them for? Well, I do it because it isn't their penny they are wasting, it's mine! I understand your frustration, because chances are, they left on three lights, an iPod stereo system with Alicia Keys and Jay Z singing about a concrete jungle, and a 21" TV with a *Simpsons* repeat episode. Frankly, I don't blame you; I still go a little crazy. Most kids think I'm grouchy! I, along with my colleagues, believe that electricity is by far the best bargain on the planet. We have no doubt many people disagree with us about that, so if you do disagree, save your breath. Let's just agree to disagree.

But think about it like this: If a gallon of milk costs about $3.00, then a 12 oz. glass of milk costs you about 29 cents. To get that glass of milk, a farmer had to buy a cow, feed the cow, provide a place for the cow to live, clean up after the cow, milk the cow, send the milk to a plant to have it processed and bottled, and then finally send the milk to the grocery store where you bought it. You then brought it home for your family to enjoy. Every step in this process costs money, and you enjoyed your fresh 12 oz. glass of milk for about 29 cents, not including your travel costs and time to go get it.

When you turn on the light in your dining room for an hour while having a spaghetti dinner with your family and collectively as a family you have three glasses of milk, then you had about 87 cents' worth of milk at dinner. Well, now let's go back to the light that is on to help you see while you have the meal with the family. To get the light on, a piece of coal needs to be dug out of a coal mine, shipped to a power plant, pulverized to look like black baby powder, and then burned in a gigantic boiler to produce steam to turn a machine that is almost certainly bigger than the size of your house to produce the electricity. The electricity is sent across a humongous web of transmission wires to an electrical substation that transforms the electrical power to a voltage low enough to run along the side of the road. It then is delivered directly to your house and the light above the table you are eating at. After you finish your dinner with the glasses of milk, you turn off the dining room light and you're out about one penny for the power. You probably dumped more financial value of milk still in the glasses down the drain while cleaning up the dishes than you spent on the light that was on during dinner.

The people providing your milk, like the Markham Dairy Farm in Constableville, New York, deserve every penny they get for it. Working on farms is, without question, hard, thankless work, and these statements just make it that much more thankless. Sorry, Mr. Scott Markham and the rest of the dairy farmers across the world. I am just attempting to make a point. Electricity is by far, bar none, the best bargain on the planet, in my mind. You are not forced to use it; simply turn it off if you don't want to enjoy the comfort and convenience it provides.

The generators which provide electricity in bulk and look like some of the pictures on the previous pages deliver electricity in units called megawatts (MW), which are equivalent to 1,000,000 watts. The unwritten ballpark guess or rule of thumb is that 1 MW is enough power to provide electricity to about 1,000 homes. A megawatt hour (MWhr) is one million watts for an entire hour. The hour it is measured in can be prorated to calculate the actual usage for the one-hour time period. For example, a 100-watt light bulb left on for a half hour works out to be 50 watts of usage, or 0.05 KWH. A generator producing 1,000 MW for half an hour generated 500 MWH.

Okay, you can relax again, as that's it for the math. If you can handle that, you should have no trouble with the rest of Part I.

One further point that will be discussed in future chapters is that electricity cannot currently be cost effectively and efficiently stored. Electricity is the ultimate example of just-in-time delivery. The generators scattered across the interconnection must respond immediately and change their output in direct proportion to changes in customer demand. The electric system is unlike a natural gas pipeline system, where the natural gas can be stored in the pipe. The more that is stored in the pipe, the higher the pressure. In a natural gas system, the gas operators on a cold morning when they know the demand will be very high will routinely put as much in the pipe up to the point short of where the pressure alarms will trigger to inform the operators to reduce the pressure. This gas in the pipe is called line pack. Unfortunately, this is not an option on the electrical power grid.

The generating plant operators, and all of the staff who work in the generating plants, have some of the most dangerous jobs in the industry. When catastrophe occurs in a generating plant, it is usually extremely destructive. If you know of a person who works at a generating plant, it would be good of you to recognize the dangerous job that person has. Although the pieces in the plant all have built-in safety systems, you need to realize that there is inherent danger due to the enormous amounts of speed and force used in the process to produce electricity. The steam made in the huge boilers to drive a normal generator is heated to temperatures and pressures that eliminate all water, so when a steam pipe breaks, you won't see anything that looks like a tea pot boiling over; instead, you might just see a piece of steel melted and a vapor cloud forming 50 feet beyond the point that something happened. Generator turbine shafts are not weighed in pounds; they are weighed in tons and are rotating at mind-boggling speeds of hundreds or even thousands of revolutions per minute. When something goes wrong in this type of an environment, it's not going to be a warm fuzzy feeling. With that said, this is a good opportunity to say thank you to anyone who works or has worked in a generation station.

CHAPTER 4

The Transmission System

THE TRANSMISSION SYSTEM IS A web of aluminum, steel and copper wire called conductors (pieces of wire meant to easily allow electricity to pass through them). The conductors are hung from porcelain insulators attached high in the air and held by enormous transmission towers. Copper wire is probably the most durable, as it is very long-lasting and is not affected significantly by Mother Nature. Aluminum conductors are also good material for conductors; however, aluminum conductors do not age as well as copper conductors. The weight of aluminum is considerably lower than that of copper, making stress on the towers less and the construction process slightly easier. Steel conductors are not as popular as aluminum or copper but are often used in combination with aluminum conductors. Steel is sometimes located in the middle of a conductor that is then wrapped with aluminum. This strengthens the lightweight aluminum conductors but still provides an excellent conductive material.

Below are a few pictures of some high-voltage transmission towers.

Source: Shutterstock.com

Source: Shutterstock.com

Source: Shutterstock.com

Source: Shutterstock.com

Source: Shutterstock.com

The transmission system is also partly located underground, at an expense of close to 10 times what overhead transmission costs to build. The additional cost is attributed to many different things. Some of them include necessary regulations and permits of building underground networks, specialized construction materials, special equipment needed, and the right-of-way construction requirements.

When the transmission system is located underground, the power travels on solid dielectric high-voltage cables, which are often surrounded by circulating oil to cool them. I realize I just used the term dielectric and won't try to explain any more than this: Dielectric cable is a conductor surrounded by an insulating material (substance that prevents electricity from passing through it—rubber, for example). If you want to see why I am trying to keep this as simple as possible, look up "dielectric" on Wikipedia; it's a good explanation but will also give you a taste of how complicated this subject is. I freely admit I don't know how it all works at a physics level; I only know how it works in reality. Transmission lines hung from the towers, called overhead transmission, have the benefit of open air to help cool the conductors. Most underground transmission systems are called networks, and they are located in metropolitan areas. A detailed explanation of underground network cable systems is not completely necessary at this stage; however, it is worthwhile and important to at least make you aware that they exist. If you pay attention next time you drive into a big city like New York City or Chicago, you will notice that as you're getting close to the city, you'll probably see lots of transmission towers heading in the direction of the city but when you get into the city itself, you will see very few or none.

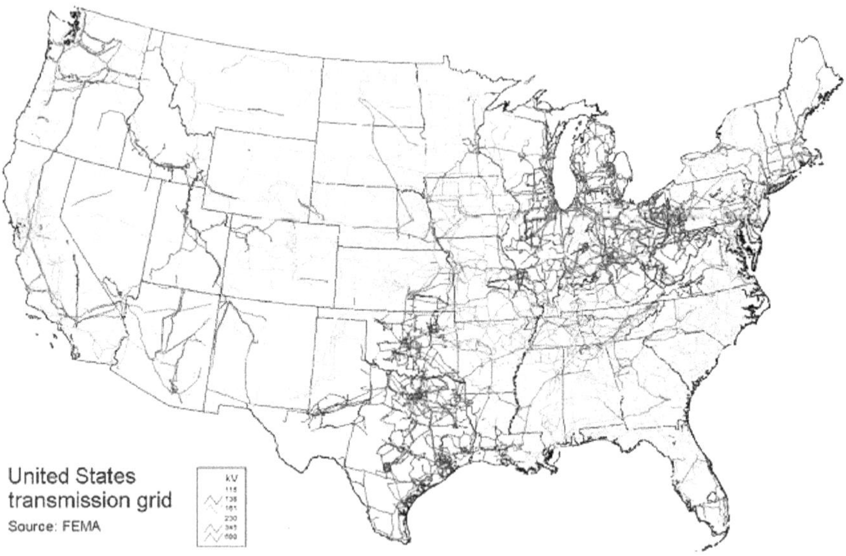

Transmission System Source: Wikipedia

This is because at some point, the electric companies took the transmission system underground. The large majority of the country's electric transmission system, however, hangs from tall towers similar to the depicted images in preceding pages.

Above is a map of the United States with a significant portion of its transmission system displayed. The lines in this map are normally colored, and it would have been nice to have them colored here as well. However, if you take the time to look at the website source listed, you will see the different colors. They represent different voltages levels for the high voltage lines. Engineers would explain the different voltage levels and use technical terms, such as impendence, resistance, and power transfer ability, and give you the reasons why they are different, but for simplicity's sake, I will ask you to think of the transmission system as a road system similar to that we have for vehicles. The higher the voltage level, the more of a major interstate highway it represents and can move large amounts of megawatts. The lower the voltage level, the more of a winding back-country road it becomes, and it can move fewer megawatts. The higher the voltage, the more MW a line can move, which in turn helps reduce the number of transmission lines needed to be built at lower voltage levels, creating fewer transmission corridors needed to deliver the energy to where it is needed. The higher the voltage level, the fewer the electrical losses in the delivery of the electricity to the next segment in the path. Yes, losses. Remember earlier we talked about heat in terms of the dielectric cables and keeping them cool? Well, the transfer of electrons along a path heat the cable or conductor, and this heat becomes the unintended byproduct. The result is a loss, the amount of electricity lost from one end of the line to the other.

If this discussion started to get over your head, just realize the transmission system's main purpose is to deliver electricity (megawatts) from the generators scattered across the system to substations and distribution systems for eventual delivery to the customer. In the upcoming section "Real-Time Operations," you will learn a bit about how the transmission system is protected from failures, how taking a line out of service for maintenance is coordinated, and how forced outages on the transmission system are responded to.

Transmission Right-of-Ways

When it comes to operation of the grid, industry experts can very easily determine where new transmission should be built in order to maximize reliability and/or profits of the power grid. That's the extremely easy part. The actual construction of the new transmission is relatively easy as well, provided the geographical landscape is easy to handle. Is that to say construction crews have an easy job building transmission lines? Absolutely not. But in the overall extensive and long process, the real construction process is straightforward. The difficulty lies in the planning process for new transmission because of right-of-way considerations. A transmission line must be free from obstructions that can cause the lines to unexpectedly go out of service. In order for lines to be free from obstructions, the lines have to clear vegetation up to a certain predetermined width for the line to go through. The voltage level the line operates to will determine the construction width of the right-of-way needed to be clear of vegetation and other obstructions. Here is an example of approximate widths for differently built transmission lines operating at different levels.[9]

Nominal Line (Voltage level)	ROW Width (feet)
69,000	50–100
115,000	75–125
138,000	100–150
161,000	100–150
230,000	150–200
345,000	150–200
500,000	175–200
765,000	175–200

Now, try to imagine the town you live in and another town approximately 50 miles away from you. Picture carving a path that is 200 feet wide connecting your town and the other, and how many homes, businesses or other obstacles might prevent the line from ever getting to the construction phase. If you're having difficulty picturing how

Transmission Right-of-Way   Source: Shutterstock.com

wide 200 feet is, think of more than half a football field, or roughly 10 times the width of a normal neighborhood road you drive on. Think of all of the people impacted and the battles they will stage to prevent the project from occurring. There is a term, NIMBY, that summarizes many of the warranted and reasonable emotions of these debates. NIMBY stands for not in my back yard. This is one of the many reasons the transmission grid does not expand as much as the real-time power grid operators wish it would. It is even difficult to upgrade a voltage level in a right-of-way. For example, if engineers wanted to take an existing 69 kilovolt (kV) line and upgrade it to a 345 kV line, it would require a wider right-of-way. This would be easier than creating a brand-new right-of-way but would still require extensive study, hearings and planning.

When media and others refer to the aging power grid, the difficulty of getting an approved right-of-way is part of what we believe is their concern. The ability to move megawatts to the areas they are needed is limited by the existing transmission system. Our world continues to grow, and electrical demand increases slightly every year, but innovative and groundbreaking ideas continue to develop for ways to meet customer needs without building more power plants or transmission lines. Let's look at just one example of what is in the beginning stages of creation.

Power grid operators of the organizations you will read about in Part II, are sometimes responsible for areas that combine the geographic areas of multiple states. Some areas include more than a dozen states. Discussions and plans are already in place with huge national chain stores—I do not know which ones, but they are big, similar to Lowe's, Home Depot or Wal-Mart. When immediate customer demand for electricity is at or near record levels, the transmission system, and the power grid as a whole, is strained to the maximum. Generators are at their maximum operating range, which we discussed earlier. In these types of situations, power grid operators will in the near future be able to send signals that will reduce power usage to the chain stores throughout the entire multi-state region. It might be several hundred or a thousand different stores from each company. All of them reducing usage at the same time would free up the availability of electricity needed on the power grid. These programs are called demand-response programs. In the existing programs and in programs being developed, the stores will receive a signal that is sent from the power grid operator to a central office for the chain. The central office then sends out instructions to each of its stores for what power to reduce. For example, the instructions might be for the stores to turn off half of their air-conditioning and/or two fluorescent bulbs in each four-tube florescent bulb light fixture. Think about it: If every Wal-Mart and Target in more than a dozen states turns off half of its lights in the same hour, it adds up to quite a bit of electricity reduction. Even if you're in the store at the time it happens, you might notice the lighting change and wonder what happened, but you probably won't think much more of it.

These types of programs only need to be implemented when the power usage on the power grid is extremely stressed, probably no more than once or twice a year for no more than a few hours each time, so with these types of programs, the need for new expensive power plants may be avoided for the limited times the extra energy will be needed. It's a win-win for everyone. There are relatively new businesses that go out and sign up a few or even hundreds of participants to be part of a demand-reduction pool. This is a growing branch of the power industry as well. It is becoming known to the industry and the public as one of the significant elements of a Smart Grid, which we will discuss in more detail later in the book.

The physical work on the transmission system is done by people with several different job titles you will recognize as lineman, electricians, and relay technicians. These individuals, like the generating plant operators, have extremely dangerous jobs. There are several safeguards built into the process to keep the field personnel safe from injury when they are taking equipment out of service for maintenance or emergency conditions. One of the many safeguards is the communication process for how the equipment is removed from service. A switching order is written by individuals in a control room and approved by one or more people reviewing it for errors. It is written to give a step-by-step process of how to get a particular piece of equipment out of service safely for maintenance. The order is then communicated over phone or radio one step at a time to people in the field. The field personnel and power grid operators go through a communication process called three-way communications (also known as three-part communication). The field personnel repeat the step back to the control room operator for clarity of understanding, and then the control operator either confirms the step or corrects the miscommunication. It is really no different than going to a fast-food burger place similar to McDonald's. Next time you place an order for a quarter pounder with cheese, medium fry and diet Coke, take notice of what occurs. More often than not, the person taking your order repeats it back to you: "Okay, I've got you down for a quarter pounder with cheese, medium fry and a diet Coke." They wait for your confirmation of some kind—a head shake or a verbal ("Yup, that's it.") verification, so even at fast-food restaurants, three-way communication is occurring. The biggest impact to a mistake in the fast-food industry is a waste of a couple of bucks and a dis-satisfied customer. The severity from mistakes occurring because of miscommunication at a fast-food restaurant is far less than a mistake on the power grid. Fatalities, injuries, millions of dollars in equipment damage, and loss of power to customers have resulted from poor communication in the power grid. It is through those experiences that changes have occurred over time throughout the industry to improve communication of important matters relating to operation of the electric grid.

CHAPTER 5

Substations

MY COLLEAGUES AND I STRUGGLED to come up with an analogy for substations for you to easily understand the reason for them. The first of many ideas was exit ramps of an interstate highway, but that does not accurately reflect enough of what substations are. Eventually, it was agreed that air travel just might do it.

Suppose you're trying to get from New York City to northern Canada for a fishing trip at the Jackson's Lodge and Outposts Fishing Camp (I have never been there but would like to go). So, you catch a commercial flight on a 747 that seats about 400 people out of JFK to O'Hare in Chicago (this is the transmission system), you then get onto an E75 that seats 70 people from O'Hare into Winnipeg, Manitoba (this is still the transmission system, but it is at a lower voltage level). You then hop on a float plane to drop you at the lakeside cabin you'll be fishing from (that's the distribution system). The dock from the plane to the shoreline of the Jackson's Lodge and Outpost act as the power lines from the telephone pole to your house. Substations represent all of the locations of all the transitions you made on the trip getting there.

Below are a few pictures of substations so you are clear what is being discussed. You have probably driven by dozens, maybe hundreds, of these types of structures in your lifetime.

Source: Shutterstock.com

Source: Shutterstock.com

Source: Shutterstock.com

Source: Shutterstock.com

Source: Shutterstock.com

As you can see, the possibility of failure exists in the equipment that is inside the fenced-in substations.

Substations have many purposes. One of the best technical explanations is written by Jack Casazza and Frank Delea in the book *Understanding Electric Power Systems.*[10] Their one- or two-line descriptions of what substations do have been included below in bold font. Their descriptions are a good summary of the many purposes of substations. My words (not in bold print) further expand on their statements, but my words are in simpler terms.

"Allow power from different generating plants to be fed into the main transmission corridors." As you saw from Chapter 3, there are many transmission lines throughout the interconnection. The MWs that the generator produces travel to where they are needed. Many more MWs are needed in Chicago, Illinois, than are needed in the village of Catskill, New York, so for efficiency, high-voltage transmission was built to bring large quantities of power to metropolitan areas. Substations provide a place to feed into the high-voltage transmission system or provide a place to drop power off of the high-voltage transmission system.

"Provide a terminus for interconnections with other systems." There are many different companies and areas embedded within an interconnection. The best way to describe the areas and not be exactly perfect would be to compare areas of the electrical grid to law enforcement throughout the states. State police officers have states they are responsible for, and they rarely cross state lines to patrol. The state lines for, State Police officers, are the boundaries where responsibility is transferred. Since the Eastern Interconnection is connected from the Florida Keys to the northern portions of Manitoba, Canada, many eyes are needed to operate it. That is one way the boundaries of ownership have developed. Substations mark the end of the line in some boundaries and the beginning in others.

"Provide a location where transformers can be connected to feed power into the sub transmission or distribution systems." Transformers are pieces of equipment common to most substations. They are used to transform, or change, the voltage from one level to another. Sometimes they are used to step the voltage up to a higher level, such as when a generator produces MWs and the MWs go through a transformer to step up the voltage to feed into the transmission system. The higher the voltage, the more MWs can travel on the lines going out of or into the substation. The electric distribution system uses lower voltages than the transmission system to send useable power to your home. This voltage, prior to entering your home, is once again stepped down using a transformer hanging on a utility pole that looks similar to a large light grey can with wires sticking out of it, or a pad mount transformer enclosed in a steel box (normally green) near your home. The process of stepping up or stepping down voltage from a transformer is not to be confused with raising or lowering voltage on the transmission system. This will get discussed briefly below.

"Allow transmission lines to be segmented to provide a degree of redundancy in the transmission paths." A transmission path is not what your brain is attempting to tell you it is. Some people hear the words "transmission path" and their minds create a vision of a transmission line and towers going across a field and into a section of woods where all of the trees have been cut down to provide a path for the transmission line to run through. The image the mind created is really of a right-of-way, which was described in the transmission systems chapter you read earlier. The term "transmission path" used by power grid operators can refer to a single line or a group of lines all originating in different locations but traveling the same direction toward a common area. For example, it could be described as a group of lines originating in different areas of upstate New York; a large percentage of the MWs on those lines are traveling from north to south to help feed the high-demand areas of New York City and Long Island. If one of those lines becomes unavailable, the redundancy of the other lines in the "path" is designed to pick up the loss so nobody's lights go out as a result. The following illustration is intended to help summarize this concept.

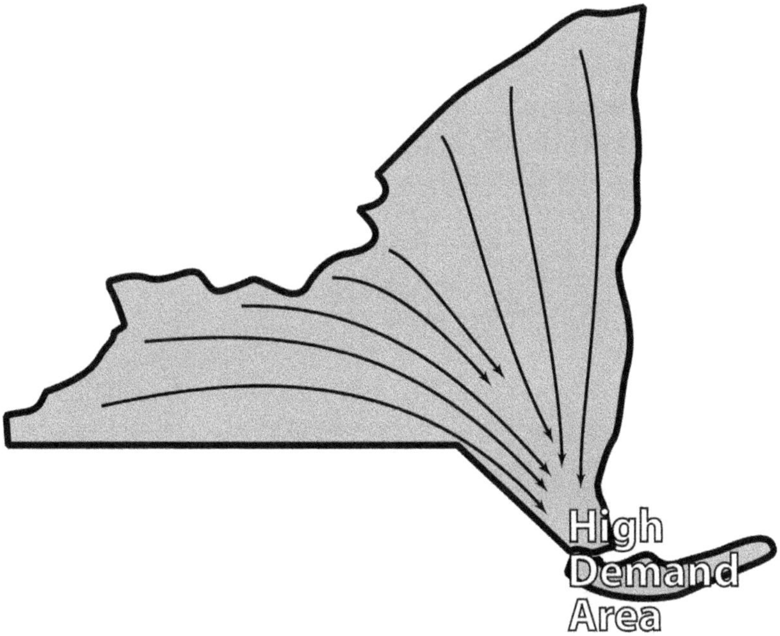

Mock diagram of New York Transmission Paths Illustration I-6

The general direction of flow for all of these lines is from anywhere on the diagram to the bottom right of the diagram, where New York City and Long Island are located. Remove one of the lines from service, and the electricity is able to find its way on other paths, or lines, to the same location.

"Provide a location where compensation devices such as shunt or series reactors or capacitors can be connected to the transmission

system." These compensation devices are also much needed and common pieces of equipment found in substations. Capacitors and reactors are used to incrementally raise (boost) or lower (buck) the voltage, which is different than stepping the voltage up or down to a new level as transformers do. Voltage is the "push" of MWs. The voltage theory is often compared to the pressure on a city water system or natural gas system. A transformer might change the voltage level from 138KV to 345KV. A capacitor might boost the voltage from 135KV to 139KV.

"Provide a location where transmission lines can be de-energized either for maintenance or because of an electrical malfunction involving the line." Maintenance of your home, your vehicle, and your furnace is no different than maintenance of equipment on the bulk power grid transmission system. It is as important as anything else. For maintenance to be provided there needs to be a place and process available to the power grid operator to be able to remove small pieces of the grid from service so that it can be maintained safely and without disruption of service to customers. Also, these are points at which, in the event of physical damage, the transmission line can be "automatically" removed from service before any further damage occurs to the equipment of the rest of the grid.

Remember when we discussed blackouts caused by cascading events? Well, some of these cascading events were caused by failures in the automatic protection systems used throughout the grid. Below is a more detailed description.

"Provide protection, control, and metering equipment." We need to break this point into three parts for a better explanation.

Protection

For operation and reliability of the grid, this function of substations is arguably the most important. The protection is really not all that much different than protection of the electrical circuits throughout your home, but on the bulk power grid, the equipment is much larger and much more expensive. Your home has several different means of protecting equipment. These may start as simply as a fuse inside an electronic device, such as a TV. Some of the outlets in your home have a component on them called a ground fault indicator (GFI outlet). The next area of protection you have is a panel box with circuit breakers or fuses to protect the individual circuits from faults or overloaded conditions.

The bulk power and distribution systems are designed the same way. When abnormal conditions occur on the power system, there is equipment in substations to help isolate faulty equipment and protect the rest of the electric grid. There are relays designed to do several different things, depending on which type of relay you are talking about, but in general terms, they all do the same thing: They send an instruction to open a breaker which creates an opening in the circuit. This protects and isolates the equipment and it is all done in fractions of a second.

Control and Metering Equipment

Real-time operators of the grid need to be able to react to the ever-changing conditions on the system. It's one thing for operators to sit in chairs in high-tech, impressive-looking control rooms, but what can they control? And how do they control it? Nearly all of the elements on the bulk power grid have some type of unit of measurement, some called MWs, some volts, some MVA, and some VARS. There are more, but that's enough. Some data points read a measurement of the status of a breaker, either open or closed. All of those data-point values total more than hundreds of thousands throughout the bigger interconnections. In the smaller interconnections of ERCOT and Quebec, those numbers are probably in the tens of thousands. All of these points are the metering required to operate the electric power grid.

That data is all routed to various control rooms throughout the interconnection, and power grid operators analyze the data with the aid of computer software. Their analysis of this data determines the actions and control they will implement. The control can be in the form of opening a breaker hundreds of miles from their office by looking at a station diagram on their computer monitor and locating the element they want to open. The operators can also determine the need to either raise or lower generation and can control this action by entering operating criteria into a software program or by contacting the generator plant operator, or a combination of both.

This is another opportunity to mention the new initiative that is emerging called "Smart Grid." Overwhelming you with information about the Smart Grid is not my intent; I'm going to wait until the later portion of the book to do that. For now, understand that the term "Smart Grid" is primarily about control and metering. It is intended to help power grid interconnections operate in a more automated manner and, at the same time, increase financial efficiency as well as reliability. Exactly what you will be talking about when the term "Smart Grid" is brought up depends on who you're talking to. An engineer could be discussing the added monitoring capability to certain areas of the power grid where readings on equipment are taken dozens of times each second and archived. It could be a steel mill businessman who now has automated equipment at the steel mill site that draws power from the transmission system. This automated equipment will determine every five minutes if the steel mill will run for the next 5 minutes or not depending on the immediate cost of electricity. A homeowner might use the term to refer to having the electric meter on their home replaced with a smart meter. These smart meters will allow homeowners to review daily electric usage in their home and how much that equates to in dollars. All of these different scenarios require control and metering equipment.

CHAPTER 6

Distribution System

THE DISTRIBUTION SYSTEM IS THE most common and/or familiar sight when people think of the electric system. To the power grid operators, the distribution system is simply considered customer demand, most often referred to as load. There is no doubt all of you reading this book are familiar with what it looks like, but here is a picture, just in case:

Electric Distribution

Below is an example of a map that lineman and electric distribution dispatchers use when customers start to call and tell the company that their lights are out. The map will be explained shortly.

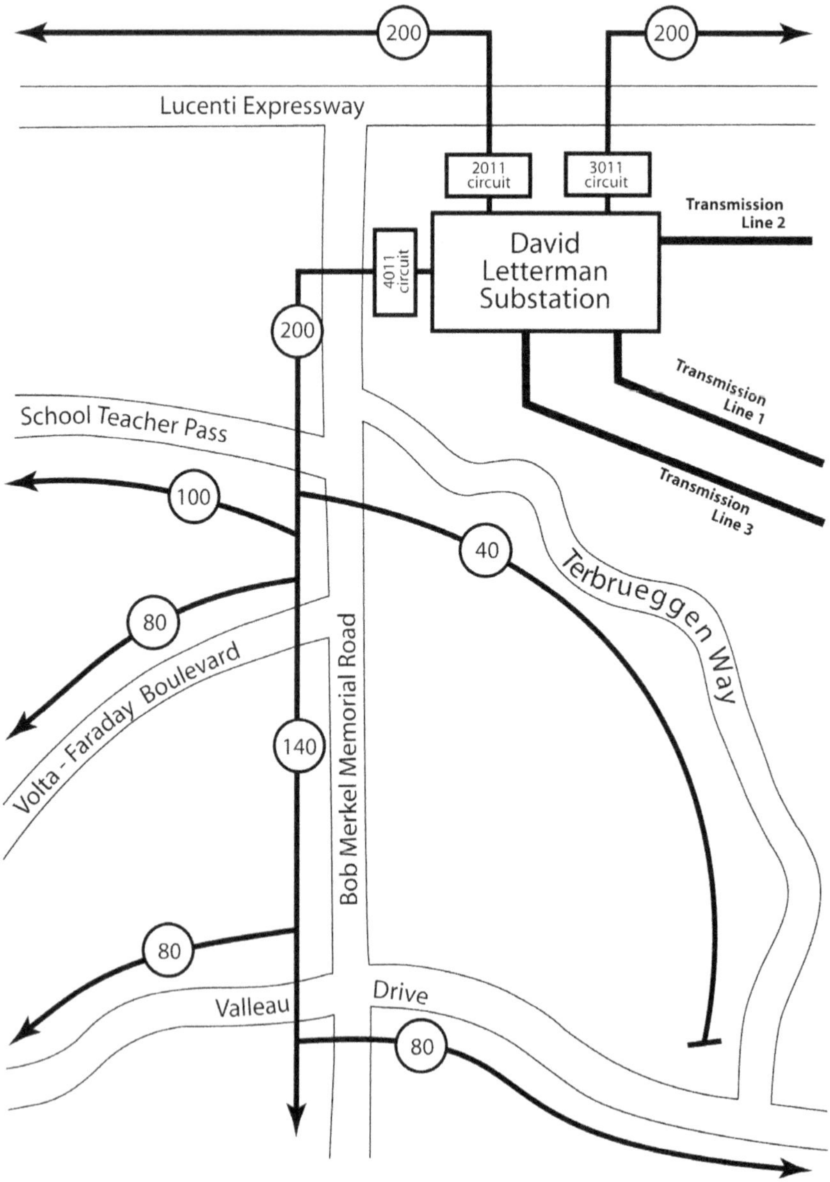

Mock Map of a Distribution System Illustration I-7

Before I explain the diagram, let's go through a short summary and explanation of the names throughout the illustration for some quick recreational reading and acknowledgement of some good people in the industry, starting with the substation:

David Letterman Substation—This is just a gross, blatant and obvious attempt to gain some national promotion of the book. This book was written on a shoestring budget, and any support to reach profitability is appreciated. David Letterman was chosen because several of us came up with a rating system to evaluate four late-night comedians. The four comedians rated were Jay Leno, Jimmy Kimmel, Conan O'Brian, and David Letterman. Letterman won the handwritten statistical evaluation process. Jimmy Kimmel came in second, with Jay Leno a very close third. Conan, well, uh, why rub salt in the wound? Let's just move on.

The three transmission lines coming into the substation are a reinforcement of a principle you have already learned: The transmission system is not radial. If one transmission line is lost, the lights don't go out. The other two transmission lines would share the loading requirements of getting enough power to the substation. If you refer back to Illustration I-4 at the end of Chapter 2, you can compare Substation A on that illustration to the David Letterman substation in this illustration. There are three transmission sources connecting the power grid to the substation and three distribution circuits feeding customer load.

The 4011, 3011, and 2011 circuits coming out of the substation have no significant meaning. They are just identification numbers so lineman, substation personnel, and power grid operators can use the data for mapping and communication purposes.

Lucenti Expressway—Ev Lucenti was a lifelong power grid operator who spent three decades performing a vast array of job functions serving needs of the electrical grid in Ontario, Canada. He was extremely knowledgeable and a true gentleman to work with on the other side of the phone. In the '90s, he became a modern-day industry pioneer to power grid operators who knew him, as he made a bold decision to leave the job and form a consulting company called Power Decisions Consulting, Inc., which is now part of the Utility Compliance Services, Inc.[11] He made this move in the late '90s as the industry was at the beginning of its "modern"-day changes. Many power grid operators watched his courageous move with envy, as his company has done very well throughout the years. I have lost touch with Ev over the years, but have thought of him often while writing this book.

School Teacher Pass—All of us through life have had opinions and beliefs in our heart and our heads, and I, for one, am no different. I have had many friends and acquaintances through the years who are educators. For much of my life I thought educators were very lucky to have such an *easy* job. For goodness's sake, they even get summers off! It took many years and a chance situation to prove my ignorance on this subject. I was involved with a volunteer job decades ago and had to deliver something to one of the local schools early in the morning. When I brought the item into the office, the principal of the school, who was a member of the same church as me, saw me and asked if I could help him out. He was in a real bind for a substitute teacher for half a day, as they could not find anyone for a morning class. He told me that he would pay me the full day rate to work just the morning, because he was absolutely desperate. I agreed to do it, as in my mind, I knew it would be the easiest money I would ever make, and it was just a few hours of my day, and heck, I was already there. Those few hours made such a lasting impression on me that I am including the day in this book as a tribute to teachers everywhere. Life offers many tribulations and heartaches along the way, but I can honestly say the four hours I spent as a teacher were some of the most mentally demanding hours I have ever encountered. My ignorance toward the education profession was forever changed by walking a mile in the shoes of educators. I will, for the rest of my life, respect and defend the grossly undervalued teaching profession.

Terbrueggen Way—Mike Terbrueggen is a man I have never met and had not heard of until I started this project. Through research for this book, I ran across his name more than a few times, and I am not sure if the nontraditional last name grabbed my attention, but whatever it was, it was enough for me to ask my colleagues if they had ever heard of him. Well, they had, and none of them had anything bad to say about him. Terbrueggen is a very well respected individual in the business. I listened to several of my colleagues' comments about him, which can be summed up in one sentence, which just happens to be on the man's website: "Mike combines 10 years of power industry engineering with 20 years of power operations training and consulting experience. He applies his power background and training capabilities to offer real world solutions to a company's training and operating needs." There are no statistics available to measure how many individuals he has helped prepare to take the required NERC certification exams but I am told it is probably well over a thousand individuals. His company is called Operations-Training-Solutions.[12]

Volta–Faraday Boulevard—A little history for you: You can't write a book about electricity and not mention Benjamin Franklin and the kite experiment. Neither can you avoid mentioning the American inventor Thomas Edison, who is holder of 1,093 US patents, making him one of the most productive inventors in our history. But if you want to sound brilliant the next time you go to your local watering hole, remember these two names: Alessandro Volta and Michael Faraday, as they had as much (or more) to do with the "invention" of electricity as anyone else. In 1800, Volta created the first ever device to produce a steady electric current. The device would later be known as the battery. And in 1831, Faraday created a device called the dynamo, which is known as the first working model version of the power generator. This invention is what began the era of electricity. In my opinion, Franklin and Edison have gotten enough credit, so I named the boulevard after Volta and Faraday, well known among engineers but virtually unknown to the rest of the general public.

Valleau Drive—James Valleau has earned a well-deserved reputation for his ability to pass his industry knowledge on to others in a welcoming-manner that few can master. He was a power grid operator for many years before moving to a more prestigious position within the organization he works for. During his years on the operations floor, he was constantly called upon to perform OTJ training of new power grid operators to the company. His humble intelligence of the goings-on related to the electrical and gas transmission systems were, and continue to be, well known throughout the Northeast and are becoming more known throughout the industry. The departmental leaders would always put the most inexperienced operators with Valleau because of his incredible ability to pass on his knowledge to the new recruit. His personality, sense of humor, attitude and operational skills have earned him a reputation as one of the best people in the industry.

Bob Merkel Memorial Road—Bob Merkel's life was cut far too short by lung cancer. He was the type of person you would meet and then say to yourself, *What a guy, I want to know him better!* There are givers and takers in this world, and Bob Merkel was a giver. He was a generous giver, at that. If you needed something—help with a project, advice, an opinion, his time, his knowledge, or his support—you got it, and you always got it unconditionally. He spent most of his career working for a small electric company in upstate New York and finished his career in the Midwest, working for one of the largest regional transmission organizations (RTOs) in the country. He was always considered one of the most respected operators in the business. We could go on and on about Bob and his wonderful family, but since the eyes are beginning to well up, I will end it here—but not without saying we all miss you, Bob!

Okay, back to the diagram in just a bit.

A large majority of the distribution system runs along the local roads we drive on and rests on top of telephone poles. It's kind of funny that most telephone poles are actually owned, installed and maintained by electric companies. Originally, before electricity was available everywhere across the country, farmhouses out in the country did have phones and the phone lines were mounted on telephone poles along the country roads. In the 1930s, President Roosevelt formed the Rural Electrification Administration to provide loans to form new organizations called electrical cooperatives. In the beginning of the 1930s, over 90% of people living in cities had electricity in their homes, but less than 10% of people living in rural areas had electricity to their homes and farms. The electric companies couldn't justify the expense of stringing the power lines through huge rural areas just to pick up a few customers. With funding from the US taxpayers, the project provided loans to allow rural areas to form cooperatives and run power lines through the countryside. This enabled them to connect to the neighboring power companies and buy power from them. This led to increased efficiency at the farms by allowing them to use mechanical assistance supplied by electricity. But anyway, the term "telephone poles" stuck, and nowadays, the phone companies and the cable TV companies often rent space on the "telephone poles" from the electric companies that own and maintain a large majority of the poles.

The main purpose of the distribution system is to connect the transmission system to the customers. Unfortunately, the measure of customer satisfaction regarding reliability in this area is misleading to bulk power reliability perception. When the lights in your home go out, it is nearly always because of a situation on the distribution system. One reason might be that an animal got into something he wished he hadn't, a car drove into a utility pole, or a tree limb fell on the line. The distribution system uses compiled statistics measuring

SAIDI—System Average Interruption Duration Index,
SAIFI—System Average Interruption Frequency Index, and
CAIDI—Customer Average Interruption Duration Index.

Utility companies use these indexes to determine how reliable the distribution system is and in what locations improvement is needed.

Power grid operators are often asked by neighbors or friends about their power unexpectedly going out "the other day" or "last night." The conversation might sound something like this: "Hey, my power went out last night. What happened? I noticed nearly everyone was out, but some of my neighbors had power. How can that be?" Hopefully, I can explain one of the most common reasons so you will understand it.

The distribution system has a form of protection called line fuses through a large percentage of it to protect it. This is what they look like:

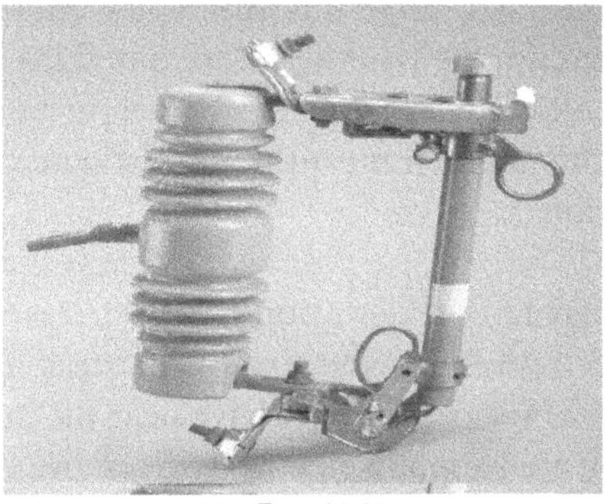

Line Fuse Holder

The circular tube is hollow, and inside, it contains a small-diameter wire known as the fuse. The fuses are designed to protect the much thicker wires that you notice as you drive on your local road. Without the fuses, the fault would be detected all the way to the substation and a breaker there would open to discontinue electric service to all of the customers on the affected circuit.

This is what it looks like on a pole when one of these fuses blows.

Two line fuse holders – one open and one closed

As you can see, in the diagram there are two line fuse holders, one is in the closed position, and the other is hanging down in the open position, as it is not connected to the top of the holder. This *hanging down* shape is what the line crews will hopefully find when they are investigating and trying to determine why your lights are out. This is the break in the circuit preventing electricity from getting to your home. What caused it to blow is what the linemen must then investigate before putting the new fuse in the holder.

The fuses are displayed in Illustration I-7 as circles with a number in them. The number represents the size of the fuse. Take a look at the map and imagine you live on Valleau Drive near the intersection of Terbrueggen Way. You are supplied electricity from the 4011 circuit out of the David Letterman substation. The 80 amp fuse at the beginning of Valleau Drive blows, because a neighbor's kid was flying his fancy kite with little control on a very windy day. So now the power goes out to the customers on the east side of Valleau Drive. But now remember where you live in this scenario. All of the customers around you on Terbrueggen Way have power because they are fed from the 40 amp fuse at the beginning of the road, up near the intersection of Bob Merkel Memorial Road. You may also notice, if you live on Valleau Drive, that your lights will go out any time the 140 fuse or the 200 fuse blows. The farther you live out on the circuit, the more opportunities for your lights to go out. So, if your lights go out more than a friend of yours claims theirs goes out, it is probably because you live near the end of the distribution circuit. There are many more scenarios, but these are very common explanations our friends and neighbors have understood over the years.

Electric distribution systems are very similar throughout the interconnections. There are some small differences in equipment and design from region to region, but in general, they are very much alike. This is the reason why companies are able to engage in a practice called mutual aid assistance. Mutual aid assistance is an agreement to help each other out in times of dire need. If one region of the country is affected by severe weather, such as a hurricane, and the destruction to the distribution system is enormous, with thousands of customers without power, the companies this happens to can ask for help from other areas of the country that were unaffected by the storm. For example, if a hurricane causes severe destruction in North and South Carolina, help can be called upon from companies in Georgia and Florida that may not have been affected as badly. It is possible the companies in Georgia and Florida have been impacted to a lesser degree so they can't let a majority of their linemen and other field personnel go to help in the Carolinas, so the companies looking for help reach out further, and other companies in the states of Virginia, Maryland, and in the Northeast can pack up their bucket trucks and head south to help. The same case is true if a major ice storm hits the New England and New York area. The companies in the southern states can head north to help restore customers. It's not done out of the goodness of their hearts; the company requesting help ends up paying the linemen and their parent company for the use of the equipment and personnel. If you happen to live in an area where a major weather incident happens, just take notice, as you just might see your normal power company truck working with a truck from somewhere hundreds of miles away. This way, they can get all of the local power back up and working as fast as possible; by having enough people to work around the clock while still letting some people have a little time to sleep.

CHAPTER 7

Legislation, Governance and Regulation of the Grids in the United States and Canada

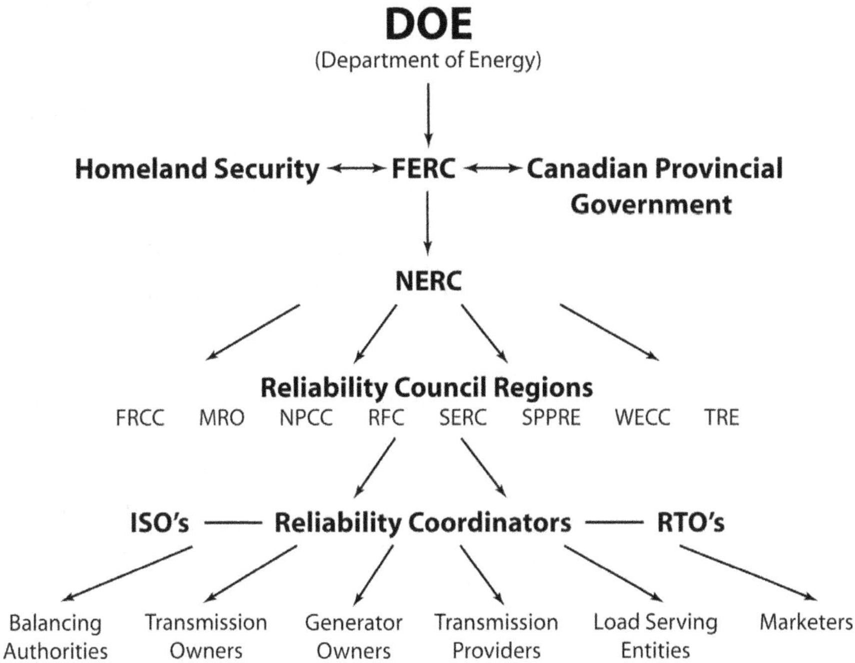

Governance of the Electric Power Grid Illustration I-8

Above is the hierarchy of the legislative governance of the power grid. Below is a brief description of each of the levels. Some levels are occasionally combined or overlap in the real world.

PUC (Public Utility Commission) or PSC (Public Service Commission)— I know, I know, they are not on the illustration. I worked with a few people on this illustration, and we all came to the same conclusion about this important part of the legislation and governance: They ABSOLUTELY belong on this diagram, as they do have significant input into several issues in the industry. Words often used to describe the PSC or PUC are power, influence, command and persuasion. The only problem was where to put them on the diagram. They affect nearly every level of governance, so the people I spoke with agreed to leave them off of the diagram but to mention them first. When I asked an electric industry leader about the diagram and where he thought they should go, his eyes lit up and he said, "I'd like to tell them where to go!" I think he was just kidding around, as he began to laugh. These organizations serve the public by keeping a watchful eye on all utility companies, including electric. Just a few of the many services they provide and keep a close watch on are cost rates charged, environmental considerations, safe and reliable service, and assistance with utility company disputes with the public.

DOE (Department of Energy)—The DOE is headed by the Secretary of Energy. This position is appointed by the President of the United States. The DOE has many responsibilities within the federal government. One of its many responsibilities is the power grid. In the Department of Energy, there is a regulatory agency called the Federal Energy Regulatory Commission (FERC) tasked with this job.

FERC (Federal Energy Regulatory Commission)—FERC has the responsibility and authority passed on by Congress to enforce rules and regulations that companies who either operate the grid or are part of the grid must follow. In other words, FERC enforces the standards by which the grid is supposed to operate. FERC also regulates the transmission and wholesale sales of bulk electricity in interstate commerce and examines and investigates oddities within energy markets. FERC lacks the detailed subject matter expertise of some of these tasks, so it relies on the North American Electric Reliability Corporation (NERC) for help in these efforts.

Homeland Security—Homeland Security is tasked with the security of our nation. One of the many, many things considered in reference to the safety and security of our country is the electrical grid, so this group has input as to the safety, security and protection of various components of the grid as a whole.

NERC (North American Electric Reliability Corporation)—The NERC is responsible for promoting the reliability of the electrical grids in North America. NERC works with stakeholders to develop the policies and standards to which the organizations and operators of the grid must adhere. Failure to operate to these predefined rules can cause companies to be subject to financial penalties that have reached the tens of millions of dollars. The rules NERC uses to promote the reliability of the bulk power grid are known to the industry as standards and can be found on the NERC website. Each of the 14 standards has dozens of requirements that the bulk power grid is to be operated under. If one of these requirements is violated, severe financial penalties can result. The 14 standards are listed below.[13]

Resource and Demand Balancing (BAL), with 101 requirements
Communications (COM), with 13 requirements
Critical Infrastructure Protection (CIP), with 172 requirements
Emergency Preparedness and Operations (EOP), with 110 requirements
Facilities Design, Connections and Maintenance (FAC), with 143 requirements
Interchange Scheduling and Coordination (INT), with 40 requirements
Interconnection Reliability Operations and Coordination (IRO), with 97 requirements
Modeling, Data and Analysis (MOD), with 243 requirements
Nuclear (NUC), with 35 requirements
Personnel Performance, Training and Qualifications (PER), with 18 requirements
Protection and Control (PRC), with 125 requirements
Transmission Operations (TOP), with 71 requirements
Transmission Planning (TPL), with 86 requirements
Voltage and Reactive (VAR), with 31 requirements

This totals 1,285 requirements for organizations involved in operating the bulk power grid.

NERC also investigates major events on the grid and performs analysis of the events to help prevent similar situations from occurring in the future. NERC has additionally instituted and developed the process by which operators of the grid are certified to be allowed to make the necessary real-time decisions to keep the grid safe and reliable. A certification program is in place, which requires that all individuals who make real-time decisions related to the grid be certified. This certification program can be compared to licensing programs for physicians, lawyers and pharmacists, among other professions.

The first step in the NERC certification process is as NERC states on its website:

> *The System Operator Certification Program awards certification credentials to those individuals who demonstrate that they have attained sufficient knowledge relating to NERC reliability standards and the basic principles of bulk power system operations by passing one of four specialty examinations:*
>
> 1. The **Reliability examination** is designed for system operators working in reliability coordinator control centers with a wide-area view.
> 2. The **Balancing, Interchange and Transmission examination** is designed for system operators working in control centers from which both generation and transmission are controlled.
> 3. The **Transmission examination** is designed for those system operators working in control centers from which transmission is controlled.
> 4. The **Balancing examination** is designed for those system operators working in control centers from which generation is controlled.[14]

Once operators pass one of the four exams required for their needs, they are not done. They must maintain their certification by taking training classes and participating in simulation exercises, and they must provide proof of attendance of these classes. There is a mandatory number of annual hours needed to maintain certification. Many companies have their own NERC-approved training programs which provide their operators with the necessary yearly credits, which are called continuing-education hours (CEH). Failure by an operator to get the required amount of hours will cause the employee to be removed from the job function. That individual will be unable to make real-time operating decisions for the company he or she represents until obtaining the necessary credits and/or taking the certification exam again.

There are several vendors which are recognized by NERC to provide CEH (training) to anyone who requires them and is willing to pay for it. Companies routinely contract with some of these specialized vendors to come to their offices and train their operators. I list a couple of them in a section near the end of this book.

NERC works with and depends on the Regional Entities for direct oversight and auditing of all of the RTOs, independent system operators (ISOs), reliability coordinators (RCs), transmission owners, generator owners, balancing authorities and others. The Regional Entities then report back to NERC the findings of those audits. The audits are designed to make sure the RTOs, ISOs, RCs and members of the grid are operating to the standards and policies approved by NERC and FERC.

NERC works eight Regional Entities for auditing services:

FRCC—Florida Reliability Coordinating Council
MRO—Midwest Reliability Organization
NPCC—Northeast Power Coordinating Council
RFC—Reliability First Corporation
SERC—SERC Reliability Corporation
SPPRE—Southwest Power Pool Regional Entity
WECC—Western Electric Coordinating Council
TRE—Texas Reliability Entity

Below is a map of the area over which each Regional Entity have auditing and oversight responsibility. These organizations do not have *operating* responsibility, just auditing and oversight.

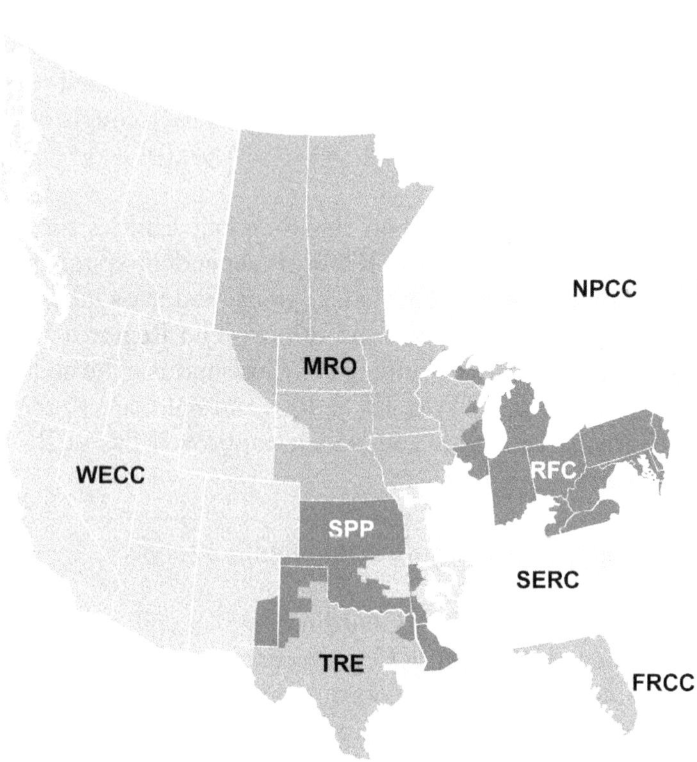

Regional Entities Illustration I-9.

RTOs and ISOs

The responsibilities and goals of RTOs and ISOs is what will be focused on in Part II of the book, but for now, here is a general understanding: These organizations were formed, in part, because of FERC's policy of providing open access to the transmission grid with the industry-changing landmark order 888/889[15] in 1996. These policies would foster energy markets, allowing generators to compete with each other to produce MWs and in turn reduce wholesale prices.

The MISO (the 2011 Franz Edelman award winner[16]) formed in December of 2001 and was the first organization in the country to become an RTO. RTOs and ISOs are independent organizations that operate a not-for-profit electricity market for their members. Members are not obligated to join but find the benefits of being members to be

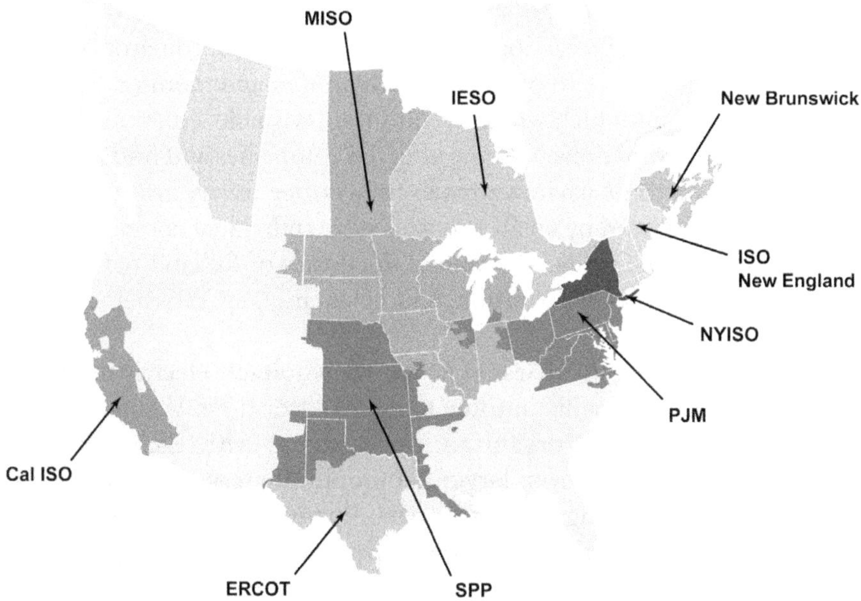

RTOs and ISOs Illustration I-10.

extremely worthwhile. MISO and PJM, as you can see from the illustration above, represent the two largest of these organizations in North America.

In the Introduction of this book, the letter to the President of the United States described the Federal Power Commission's vision for the future. The evolution of RTOs is a result of that vision nearly 50 years later. Currently, the ten RTOs and ISOs represented in the above map are responsible for about 60% of the United States' electric power grid. The other 40% of the grid is managed in the traditional method of integrated utilities and has approximately another 100 control centers managing the safe, reliable balance of generation to customer demand in addition to maintaining safe, reliable transmission paths for the energy to travel on. This 40% being referred to is in the picture above in non-highlighted areas, in the Southeast United States and much of the Western Interconnection. This is where a large number of electric companies reside but do not belong as part of a large energy market; however, these areas are still responsible for safe and reliable operation of the power grid and are still bound to operate by the standards of NERC. What they have to assist in looking at the "big picture" are entities called Reliability Coordinators. Reliability Coordinators have a staff that communicate to operators in their area and monitor data for potential problems which would result in unfavorable situations. They then discuss the situations with the affected companies and find the best course of action to fix whatever issues they notice. Every area in North America is monitored by an RC, as well as an individual company. The large RTOs and ISOs have combined the duties of Reliability Coordination, efficient market operation, and balancing generation and load.

It is the opinion of many that RTOs of huge wholesale electric markets, like MISO and PJM, will continue to evolve over time. We personally envision some existing organizations to merge with each other to encompass and oversee even larger geographical areas in the future. A global energy consulting and software company, ECCO International, Inc. (www.eccointl.com), is currently in the driver's seat to help organizations accomplish the necessary steps to create merge, or join an ISO or RTO. ECCO International has been specializing in the consulting of engineering and market economics for the energy industry for many years now and is at the top of the list as one of the proven leaders. This consulting company is well positioned and will almost certainly benefit in the years to come as more and more electric utilities transition from traditional vertically integrated utilities to become members of a competitive market structure. There is just far too much money to save to continue in the traditional manner. As I mentioned in the very beginning of this book, President Obama said, "We have a power grid that wastes too much energy." This is one of the areas that, if changed, could dramatically help reduce the waste we believe he refers to.

Those companies who choose to become members of RTOs or ISOs benefit from two primary services. The first service provided is the reliability coordination services of the electric power grid, which the industry breaks down in two different categories:

1) Power grid operators must plan and provide for enough energy to meet customer demand at all times. This is sometimes referred to in the industry as resource adequacy, which basically means that there are enough generation *resources* to *adequately* supply customer demand.

2) The operational response and recognition to the unexpected loss (called contingencies) of different bulk electric components (generators, transmission lines, transformers, etc.) on the bulk power grid—which means making sure the grid is secure enough to handle an unexpected event at all times, 24 hours a day, 7 days a week, 365 days a year. The "unexpected" is predefined to avoid confusion during the after-the-fact event reporting and analysis.

The second service that RTOs or ISOs provide is operating a not-for-profit market for electricity so the lowest cost generation mix will accomplish the two tasks of reliability listed above. This allows all of the members at the bottom of illustration I-8 to benefit from substantial financial savings that the RTO or ISO provides. The members then pass those savings on to their customers, which include you and me.

Now, how do power grid operators get the job done? Welcome to Part II.

PART II

Real-Time Operations

You turn on a light switch, and the light comes on every time. Well, if it doesn't come on, you experience some initial thoughts or emotions like surprise, shock and confusion. Why didn't it come on? Now you silently get frustrated with the company you pay your electric bill to. Then you start thinking that you should call them and inform them your power is out. "But I really don't want to go through the automated phone processes and, on top of that, where the heck did I put that number? I haven't had to call it in so long." Then you might even start to think, "Ahh what's the point? It's raining out there. . . . Probably lots of people out of lights. They probably know already that my power is off. Wait a minute! I don't remember paying my bill. Uh, yes I did. No, I didn't; no, wait, I am sure I did." All that may happen when your lights are out because most likely, something occurred on the electric distribution circuit feeding you and your neighbors. Probably not once do you ever think the entire power grid is down.

Well, now you will learn a bit of what goes on every time the light does come on, to give you good reason to never wonder if the entire power grid is down, because statistically, if you are wondering if the bulk power grid is down, then you are wasting time. This part of the book will explain the roles of the many, many power grid operators located in many areas throughout the grid.

It starts like this . . .

First comes a little planning, or maybe it is better to say A WHOLE LOT OF PLANNING. Something with as much danger, financial impact and importance as electricity needs a lot of planning. Utility companies, along with huge RTOs, ISOs and RCs, have to work together, share information and ensure that what one region plans doesn't negatively impact the power grid in an adjacent region. Some of the important things companies must consider are things like what transmission lines can be worked on tomorrow, next week or next month. Companies think about this because they need to have their crews work on these lines while they are out of service for obvious safety considerations. Taking high-voltage transmission lines out of service has an impact on the power grid, so it takes coordination with others to choose agreeable dates and times to schedule the lines out of service. Generator owners also like to plan when they can schedule generators out of service so they can perform some planned, or even unplanned, maintenance to make them more efficient or safe, or whatever the case may be. These outages must also be coordinated so there is no reliability impact to the bulk power grid.

Another thing that needs to be planned is really such a simple thing you might not even think of it if you didn't work in the industry: How much electricity will the grid need, where will it come from, and can the electricity get delivered to where it's needed? I sincerely hope you're ready to find out how all of these things are done and fit together, because here we go. . .

Load Forecasting

HOW MUCH ELECTRICITY IS THE bulk power grid going to need? This is what power grid operators call load forecasting. Large RTOs and ISOs operate electricity markets for huge—HUGE—geographical areas. Some are so big they have 13 states, some have tens of thousands of square miles, and some contain multiple gigantic metropolitan cities in the areas they are responsible for, so the first thing the organizations that operate these enormous areas need to think about is how much electricity all of their customers/members will need next year, next month, next week, tomorrow, in the coming hour, and, finally, five minutes from now. These organizations must predict how much electric usage there is going to be in the different timeframes mentioned. Right now, this may seem like a slightly strange concept to you. How could anybody possibly forecast how much electricity will be used in that many timeframes in these massive geographical areas?

These organizations have full-time staffs that do just that. They forecast demand levels for the next five minutes, the next hour, the next day, and five days out, even 30 days out. There are planners who forecast how much they'll need for the entire next year and even five to 10 years from now. The pattern of electric demand is actually relatively predictable. Mankind is, in general, a creature of habit, exercising very predictable behavior on a large scale. Weekday mornings, the demand for electricity increases as alarms clocks start going off so families can get kids off to school and get to work. Businesses, factories, schools and homeowners turn on their lights, fire up their computers, turn the heat back up, and do the things they do every other day. The things that need a little more consideration in this load-forecasting process are things like the weather and time of the year (will people be using heaters, or will they be using air-conditioners?). On extremely cold days, heaters will need to run more often than on just slightly cold days, but much of the heat is produced by oil or gas furnaces, so the electricity needed is used to just run the burners or fans that blow the heated air throughout the living or working spaces of the buildings. Air-conditioning, however, is run solely by electricity, including the compressors and heat exchangers, in addition to the fans to blow the cooled air throughout the buildings. So, although some markets which operate in far northern regions may have their annual highest usage, or peak load, in the winter, most markets in North America have their annual peak in the summer.

The company personnel that have the job of predicting load sort through historical data, looking for a day in the past which has similar data for the next 24-hour period. Some of the data they evaluate are temperature, humidity, cloud cover, wind conditions, and day of the week. It must also be in the same general yearly time frame as the next day. They make slight adjustments to account for any differences that their research exposes and give their final prediction for tomorrow's expected load for each 24-hour period . . . and the next day, and the next day, and the next.

As time moves forward and approaches the hour the forecasters tried to predict, the forecasters and the power grid operators' work together to make more adjustments to the short-term anticipated load needed. As it gets closer and closer, as you can imagine, lesser and lesser adjustments are needed, so while the load forecasting personnel may not be as accurate as they initially hoped for 30 days out, their predictions usually get better and better until the last five minutes before the initial prediction from 30 days ago. However, if a cold front moves in a few hours faster than forecasted, or if thunderstorms develop over major metropolitan areas, reducing the temperature, which in turn reduces the need for air-conditioning demand, the forecast can still be off significantly. These are just a couple of the many moving targets which power grid operators and load-forecasting teams have to consider. The power grid operators must have plans to cover these discrepancies in case the load forecast is inaccurate. Although weather is the most common cause for inaccurate load predictions, there are others. There are some things that can also affect loads and load forecasts that we can almost guarantee most people would never think about.

Tiptoe Through The Tulips

The first big media event affecting the electric grid that is remembered was when Tiny Tim married Miss Vickie on *The Tonight Show* with Johnny Carson in 1969. If you're under 50, ask your parents about it. The event was a big media hype, and everyone stayed up to see it. As soon as Tiny Tim got done singing "Tiptoe through the Tulips," everyone across the country shut off their TVs and went to bed. It caught the power grid operators at the time totally off guard because it caused the balance between generation and load to be mismatched. This caused the frequency (which is an industry term explained in detail in a few pages) on the system to go extremely high. Power grid operators learned from this event to watch for these media-hyped happenings and to try to plan accordingly; however, it is still difficult to predict something with a large degree of accuracy when you're not accustomed to it, because it has no predictable historical pattern to learn from.

The Last M*A*S*H Episode

*M*A*S*H* was a half-hour television series that ran for 11 years. It was a sitcom about a team of doctors and nurses in the Korean War. On February 28, 1983, the final episode of the show was aired, and it became the most watched television episode in US television history at the time. More than 125 million television sets tuned in to watch the show in the evening. This was not anticipated in any form by the power grid operators. Power grid operators saw a very noticeable effect on the power grid and had to make many adjustments to meet the unexpected increase in demand this show created.

The Verdict in the OJ Trial

This had a significantly noticeable effect to power grid operators as well. People in offices and homes across the entire country gradually, throughout the morning and early afternoon, tuned televisions in to await and watch as the verdict was handed down. That was not the problem. Once the verdict was announced, millions of TVs began to power off as everyone in their workplaces held their own discussions on the verdict rather than listen to the talking heads on TV. It was widely publicized by the media that companies and businesses throughout the

United States allowed their employees to watch the verdict on office TVs to help satisfy the country's curiosity. This caused another load to generation imbalance which required quick actions by power grid operators to resolve.

9/11 Terrorist Attacks

Although there was a loss of power consumption when the Twin Towers came down, the effect on the grid was minimal at best. The total loss of power being used was slightly more than 200 MW. This can be compared to a thunderstorm moving through a territory and causing several outages on the distribution system. People tuned televisions in all morning as the news spread, but the news spread over the course of hours and the population gradually tuned in to see the tragedy, so it was not a sudden impact to the grid; it was a gradual impact that was easily handled. This horrific and life-changing event had a minimal (at best) impact to the bulk power grid. I included it here because so many people have asked me about it over the years.

Super Bowls

If anyone says the Super Bowl is not the biggest sporting event in the United States, do not believe them; they are absolutely 100% dead wrong. When halftime hits, power grid operators see the load increase as people turn on lights in kitchens, bathrooms and bedrooms. People turn on lights in garages for trips to the beer fridges and/or for a quick smoke. People open their refrigerators and freezers, causing the cool air to escape and requiring the compressors to kick on to replace it. People turn their microwaves, deep fryers, ovens and stoves on for all of the goodies. They flush toilets all across the country at reasonably the same time. This then forces well pumps and city water pumps and sewage pumps to kick on across the country. There are no offsetting electrical activities, as people are leaving all the TVs on to watch the great commercials and halftime show. (Do you understand the offsetting concept? What is happening at the same time that would reduce energy usage? The answer is basically nothing. The lack of offsetting customer demand is a reasonably important concept to understand for the upcoming section of Christmas lights.) Okay, back to the Super Bowl . . .

The increase in demand on the power grid is very noticeable during this timeframe, so power grid operators need to be aware and prepared to increase generation in their areas to accommodate the increase in demand. It's a good excuse to convince all of the industry bosses that the power grid operators actually need to watch the game while working to prepare for these moments. As soon as the game is over, the load decreases rapidly as millions of people turn off TV sets and their lights and head for bed, but even this is not always predictable. Some of you might remember the 2008 Super Bowl. It was the year Eli Manning and the New York Giants prevented the New England Patriots from having a perfect 19-0 season by winning the Super Bowl. The game went down to the very end, and the New York Giants pulled off one of the biggest upsets in Super Bowl history. As the game ended, power grid operators anticipated the predictable behavior of millions of people shutting the TVs and lights off and going to bed, but something very different occurred. Not only did demand not reduce, but the demand for electricity actually went considerably higher. The predictable behavior of going to bed was disrupted by the excitement of the game, and many people left their TVs on to see who got the MVP award and to watch postgame interviews, so while the "husband" stayed downstairs to continue watching the postgame, the "wife" went upstairs to brush her teeth and get ready for bed, and she began to think, *Heck, why not flip the bedroom TV on so I can see that cute Eli Manning be interviewed?* Truth is, we will never know for sure what caused the increase in demand; it could have been that everyone in the country except New England fans decided to have one more beer and celebrate the demise of the New England Patriots' perfect season, but the bottom line is the demand for electricity was considerably higher for the 15–20 minutes following the end of the game across the entire country than power grid operators anticipated. No other sporting event comes close to the same effect as a Super Bowl on the power grid. It's not even close. You can bet your last dollar on that.

Christmas Lights

This one here is going to take some time and thought on your part, so just a warning: Read it now, and if you don't understand it, come back and read it again when you're done with the book, as it might make more sense to you then. Sometime in November, what do millions of homeowners across the country do? They go dig their Christmas lights out of their attics and begin a short period of frustration with untangling lights and checking bulbs to get the strings to work, then they hang them inside and outside of the house. The scene from the movie *Christmas Vacation* where Clark (Chevy Chase) hands the tangled-up mess of lights to his kid comes to mind, because no matter how careful you are when putting those lights away, they seem to look like that when you dig them out a year later.

The organizations that operate these enormous geographic areas, such as MISO or PJM, operate a market that serves a significantly large percentage of the Eastern Interconnection. When everyone has their Christmas lights up, the power grid operators in these areas can see as much as 3000 MW of additional load show up across their areas. Combine just these two areas together and it's approximately 6000 MW or more. That's six billion watts of "extra" electricity needed for this lighting every night. That's enough additional electricity to provide power to six million homes. That's equivalent to six of the largest nuclear power plants in the country operating at maximum output just to light up everyone's Christmas lights. But it is going to happen every year, so it should be easily planned for, correct? Wrong. It's all about predictability and efficiency.

During many years, the lights go up pretty gradually throughout late November and early December. Homeowners find a weekend or an afternoon in which they take half a day off work to run errands and have some free time, so they decide to put the lights up and get the job done. To the power grid operators, this means the putting up of Christmas lights across their region is a nice, slow and smooth daily progression of additional demand to power them. No problem. But here is one possible twist that occasionally occurs to disrupt the lights going up gradually: If there's nice weather on the four-day Thanksgiving weekend, millions of homeowners go out and hang their lights up during that short timeframe. Now, before we go much further explaining Christmas lights, you need to understand just a small amount of historical, predictable load patterns on the bulk power grid, so I am going to jump around a little explaining this. I apologize in advance.

Late afternoon into the early evening on Sundays during the late fall and winter are periods of time that power grids see the steepest increase in load change from one hour to the next. This is the time of year where that change is commonly the largest from one hour to the next.

I am not saying they are the times with the most load (demand), so don't get confused by that. I am saying this Sunday-evening timeframe often causes the largest load *change* from one hour to the next during the late fall and winter season. It lasts for about three hours, from approximately 4:30 p.m. to about 7:30 p.m. During that timeframe, it is either getting dark or is already dark at that time of year.

Just a bit more explanation might help. On a normal weekday and even on most Saturdays, businesses and industries are open for normal working hours. As they close up shop and the workers go home, there is a staggered shutdown of electricity usage from the businesses at the same time other people are getting home and turning on their lights and heat and their stoves and ovens for dinner, so the load curve takes a noticeable but predicted increase. But remember the first paragraph of the Super Bowl section about offsetting demand? Well, on Sunday evenings, far fewer businesses are shutting down for the day to offset the demand, because on Sunday, many more businesses have been closed all day, as compared to the other days of the week. Now Sunday evening comes and all of the house lights and streetlights and stoves and additional heat (because it got dark and cold) require electricity. But now keep in mind the Christmas lights, because they all start coming on as well. But there's no offsetting decrease in industrial and business loads shutting down and discontinuing their electrical needs on Sundays. Again, they, unlike on the other days of the week, have been closed all day, so the Sunday load pattern is very different than on the other days of the week. Now, on to the point . . .

So, the weekend before Thanksgiving, the larger organizations have their power grid operators preparing and reacting to approximately 10,000 MW of load increase on Sunday evening, but if it's a nice Thanksgiving weekend, the Sunday following Thanksgiving is the difficult day to plan operations because of what you just read. These organizations could possibly see 13,000 MW of increase over the exact same time period. Truth is, they just don't know how much the weather affected the population putting up Christmas lights. It's a tremendously difficult guessing game.

You might be thinking right about now, *Big deal . . . who cares*? Well, here's the catch: These giant organizations are required to have enough generation (supply) running to make sure the needs of the grid (demand) are met, but there's another, very important side to this. In order to not be wasting money on the grid, power grid operators in RTOs and ISOs also run a market for electricity. This also means they need to be fiscally responsible and make sure they do *not* have too much generation on in order to run an efficient market. If there is too much generation committed, they're basically wasting money. In one week's time these organizations might need to make sure they have the equivalent of several nuclear power plants online more than they put on just the previous week, but if they guess wrong on how much load is needed, they will be wasting your money and President Obama's State of the Union comments will be accurate. These organizations work very hard to not waste their members' money. There are very few times like the one described above when a "little extra" will be committed to prevent not having enough on, but these times are extremely rare. It is a better-safe-than-sorry scenario. For the parts of the power grid these market organizations are responsible for, I respectfully disagree with President Obama, as there is very little energy wasted in the current setup of these organizations.

Inside the Control Room, We Begin with Some Building Blocks

OKAY, CHRISTMAS LIGHTS AND TV shows may have been a reminiscent trip down memory lane, but it is time to press forward. For the next several pages, the reading is going to transition from recreational reading to more hands-on/involved thinking and reading. Everything is written to allow the average person to understand it, so if you read it with some patience, you should have no trouble, but if at this moment you are limited on time or feeling a little lazy, put the bookmark in this page and come back to it when you're in a different frame of mind.

The electricity markets that operate throughout North America all run on a very similar premise: Commit enough generation to meet anticipated demand, without needlessly running too much generation, as I stated earlier. Over the next several pages, you will slowly learn how the power grid operators evaluate the tremendous amount of data that is transmitted to them every few seconds.

A great deal about load forecasting was discussed above, but what has not yet been explained is how power grid operators know what *the actual load* (not the forecasted load) is. It is really quite simple if you don't overcomplicate it. Let's pretend you, the reader, are a power grid operator and you are responsible for the generators and transmission grid in a make-believe area of the southern half of Illinois called South Illumini, LLC. This is a fairly simple concept to explain, but it is critical to understand; it's only a page or two. It's not hard, but without knowing this, everything else will be much harder to understand. If you learn only one thing in this book, let it be this concept.

Okay, take a look at the area I just assigned to you:

Illustration II-1

There you are in the Midwest part of the United States. Your area is called South Illumini, LLC, and you have some fictitious neighboring areas called Iowatts, Inc.; Mizzu Energy; Illini Power; Hoosier Power Corp; and Kentucky Cat-a-Power. So far so good, right? Let's add to the picture some generation and transmission.

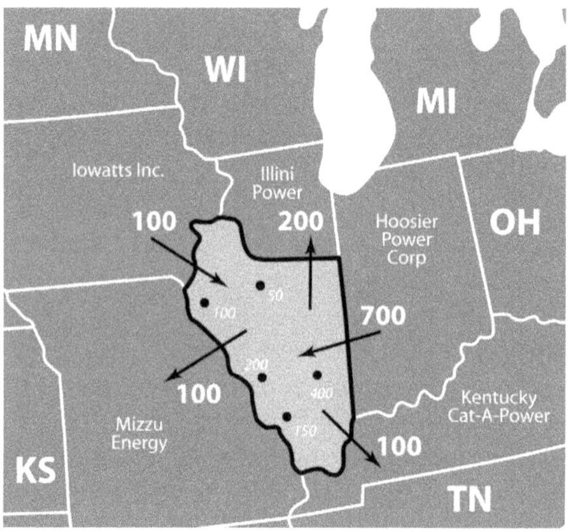

Illustration II-2

You can see there are transmission lines which connect your area and your neighbors' areas together, and these are called tie-lines. The numbers on the lines represent the flow, which is how many MWs are on the line, and the arrows show which way the MWs are traveling. The tie-lines are either exporting from your system or importing into your system, depending on the direction of the arrow. The generators in your area are represented with dots and have numbers in italic font near them. Those numbers reflect the actual MWs being produced at exactly the same moment in time that the transmission data was displayed. The numbers are updated on the computer monitors every few seconds. Your South Illumini area may have more generation capacity available, but in this example, it is not needed, so those generators are offline, and not included on this illustration.

So far so good? Now it is simple math. Add up all the MWs being generated in your area (starting from the top: 50 + 100 + 200 + 400 + 150). You come up with 900 MW, right? Good. If not, oh boy, there is no hope for you. Might I suggest you go get a Dr. Seuss book instead? Now add the flows on the transmission tie-lines together, but keep in mind whether the flow on them is leaving your area or coming into your area:

Iowatts, Inc.	100 MW into your system
Hoosier Power Corp.	700 MW into your system
Illini Power	200 MW out of your system
Kentucky Cat-a-Power	100 MW out of your system
Mizzu Energy	100 MW out of your system

Hopefully, you came up with a net of 400 MW coming into your area, so, you had 900 MW of internal generation plus 400 MW on the transmission system entering your area. You have 1300 MW of load in this area (900 local supply PLUS 400 net on your tie-lines). That's it; no tricks. It really is that simple.

Now, there is just one small thing to add. In the previous examples of your control area, we showed just one line tying you and one of your neighbors together. Well, chances are there is more than one line tying you together. We showed a total of five tie-lines tying you with five neighbors, one for each neighbor. The reality of it is, power grid operators use only one value like is shown in the example, but there are multiple lines tying you with each of your neighbors, so the map, in reality, looks more like this:

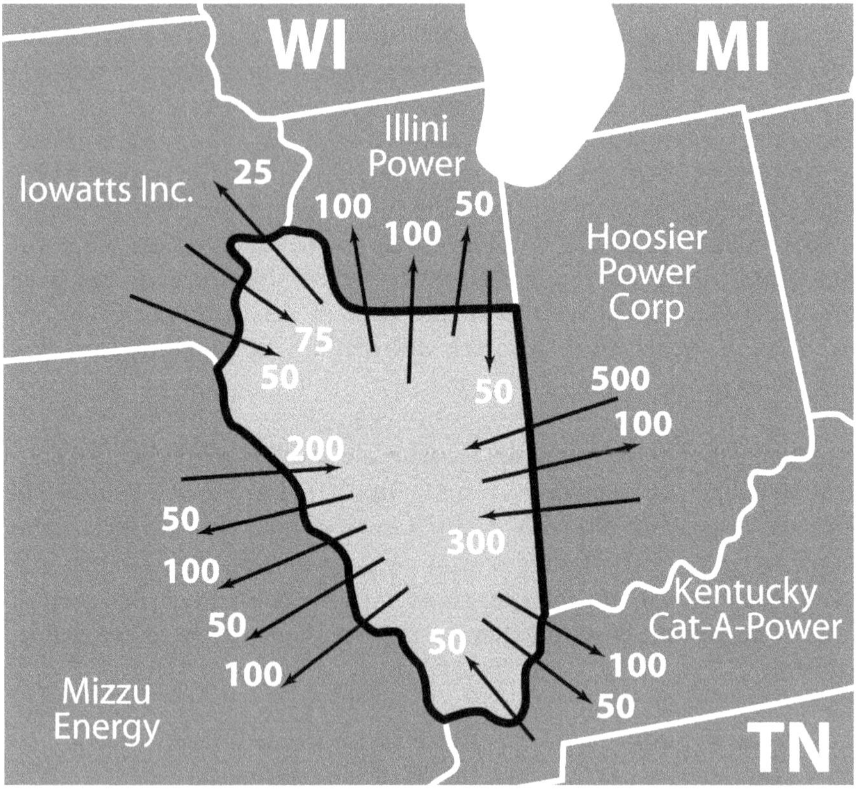

Illustration II-3

It looks daunting, but have some trust—it's not. Take Hoosier Power Corp. and its ties to South Illumini LLC. There are 3 ties:

> Line 1 has 500 MW coming into South Illumini, LLC.
> Line 2 has 100 MW going to Hoosier Power Corp.
> Line 3 has 300 MW coming into South Illumini, LLC

The sum of those ins and outs is 700 MWs into South Illumini, LLC, which is exactly what was shown in Illustration II-2 with only one value representing the total flow between the two areas.

Let's do another one from the same illustration to make sure you understand. Let's use the ties you have with Mizzu Energy, moving from north to south:

Line 1 has 200 MW going from Mizzu Energy to South Illumini, LLC.
Line 2 has 50 MW going from South Illumini, LLC, to Mizzu Energy.
Line 3 has 100 MW going from South Illumini, LLC, to Mizzu Energy.
Line 4 has 50 MW going from South Illumini, LLC, to Mizzu Energy.
Line 5 has 100 MW going from South Illumini, LLC, to Mizzu Energy.

Total the 5 tie-lines up, considering ins and outs, and you should come up with 100 MW going from South Illumini, LLC, to Mizzu Energy. How about that? That is exactly what was in Illustration II-2. You can do the same for the other companies as well. Total up the Iowatts, Inc., and the Illini Power and Kentucky Cat-A-Power. These values are something known to operators as "actual net interchange" with each individual neighbor.

So far so good? To continue building your knowledge, you need to be aware of the next simple step, which is the actual net interchange of your *entire* South Illumini, LLC. It is simply summing all of the individual neighbors' actual net interchanges with your area, like was done a few pages earlier in Illustration II-2. So, it started easy, became more complicated, and then went back to easy. That is really the entire premise of power grid operations. The data has to be immediately understandable to the power grid operators. It is much easier to look at one number than at the 18 values that the one number sums together. It's 400 MW entering your system, which totals the 18 lines. But if power grid operators need to know further details, they are summed up by individual neighbors as well. Your South Illumini, LLC, has a total of 18 tie-lines connecting you to your neighbors, but it is all summed up in one number. Why is the number 400 MW of actual interchange into the South Illumini, LLC? What would cause the 400 MW number to change?

You were told a couple of pages ago that if you were going to learn one thing from this book, this should be the concept you learn. You are very near the end of this concept—just a couple more paragraphs. So, why would the 400 MW actual net interchange value be different in the next computer scan? There are basically two possible reasons: A change in customer demand (load) inside your area is the first. The second possibility is a change in generation output from inside your control area. Below are examples for these two scenarios.

The first: Assume your largest generator—the one that is currently generating 400 MW on the east side of your South Illumini, LLC, area—develops a mechanical problem and the generator operator picks up the phone and informs you the unit is going to start ramping down to 350 MW immediately. Now, let's assume you do absolutely nothing in your chair, sitting in the control room after you hang up the phone. What occurs is the 18 tie-lines connecting your system to your neighbors all change just a little bit. The ones that were exporting will export a little less, and the ones that were importing will import a little more. Power grid operators would monitor the individual summed totals with neighbors. If you were sitting in the chair, it is quite likely you would see these values below once the generator completed its ramp down from 400 MW to 350 MW.

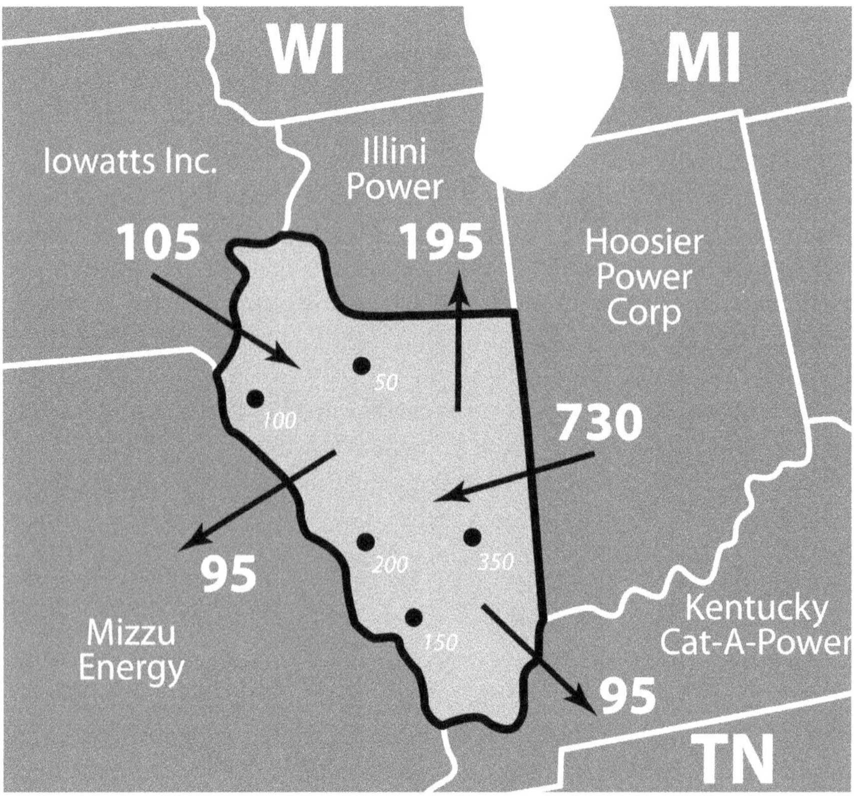

Illustration II-4

Now total up your tie-line flow. You should have come up with 450 MW entering the South Illumini, LLC, area–an increase of 50 MW into your system. This 50 MW increase offset the 50 MW decrease that was required by the largest generator in your area. The load, or customer demand, did not change at all; it remains at 1300 MW, made up of 450 MW entering your area from the transmission system added to the 850 MW of generation in your area. The same would apply, but only in the opposite direction, if your largest generator increased 50MW: The transmission flow entering your system would reduce by 50 MW, so the total would be generation 950 MW and transmission flow into South Illumini LLC 350MW.

Here is a different scenario to stress the same principles.

Assume the exact same starting point, the one from Illustration II-2, and you get an alarm that one of the tie-lines from Mizzu Energy to South Illumini, LLC, goes unexpectedly out of service. This is the tie-line in illustration II-3 that has 200 MW flowing into your area that I referred earlier to as Line 1. You immediately find out a hot air balloon pilot lost control of his balloon and ended up in the conductor between two huge transmission towers. What happens to your customer load with this line out of service? Nothing; it remains at 1300 MW. What happens to your generation? Nothing; it remains at 900 MW. What happens to your tie-line actual net interchange? Well, several of you probably said it changes. Sorry, but you are wrong. What will happen is the remaining 17 tie-lines change to make up the flow that was on the outaged line. So after the hot air balloonist has one of his worst days ever—we are assuming he is still alive but he is hanging up in the lines and he needs a change of underwear desperately—he is being as perfectly still as he can and is patiently waiting for the electric line crews to come and save him. Nevertheless, the system you're monitoring from your control room chair will change to look similar to this:

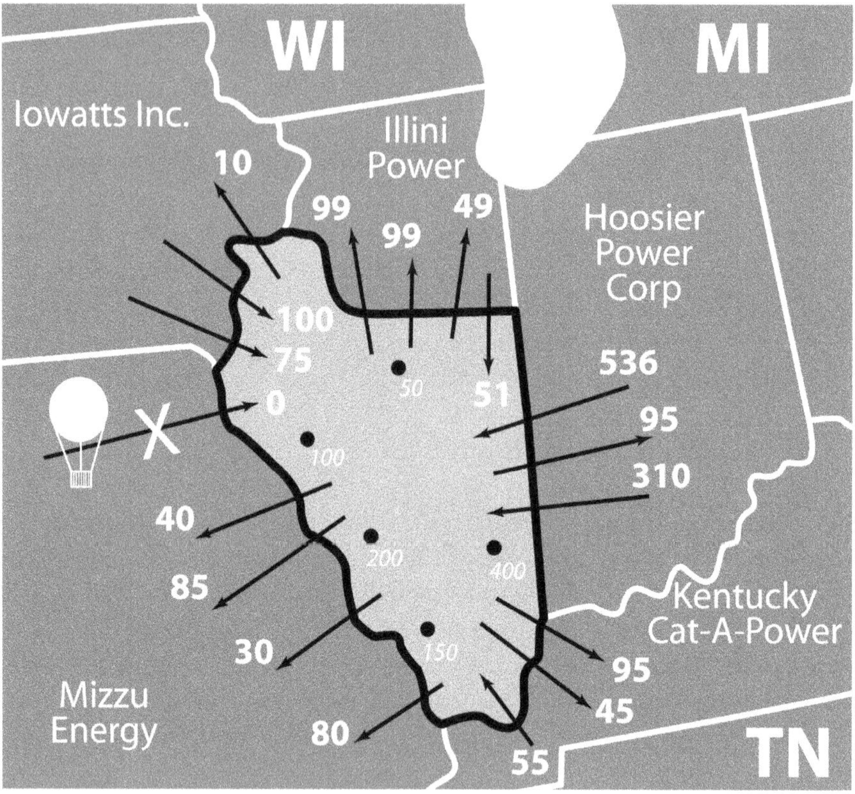

Illustration II-5

Add up all the ins and outs on your 17 remaining tie-lines, and you should come up with 400 MW. The flow on them all changed a little bit to make up for the loss of the line with the hot air balloon in it.

Frequency

Power grid operators look at a value called frequency to determine the overall health of the interconnection. As mentioned earlier, the power grid must have an equal amount of generation to load at all times. This is true, but it has a small amount of room for error, which is what we plan to explain in this section. The small amount of room for error is represented in the frequency value. In an earlier section, there was a reference to the Indy race car to operating range for generating units. Well, we can use Indy race cars here as well, , but this time we use the entire track. If the cars are all going too fast, they could end up in the wall, and if they suddenly slow down, they can get rear-ended. But if all of the drivers are driving the cars with predictable behavior, they will go around the track many times without incident. It is the unpredictable events that cause the health of the track and/or the health of the interconnection to have accidents or events. Frequency is the measurement that alerts power grid operators throughout the entire interconnection, all at the same time, that something is abnormal.

The four North American power grids run on alternating current, or AC, power, 60 hertz (Hz) systems. Standard generators connected to the grid have their generator shafts spin completely around and pass by the positive and negative magnetic poles on the rotor at exactly the same time. Huh? This is going to take a little effort on your part. It appears very confusing and technical, but at the end it's just very simple math. If a generator is designed to be a two-pole generator, the shaft will spin 3600 revolutions per minute, passing by each pole 60 times each second. That's fast . . . really fast, and there are lots of generators throughout the world that do this. There are many other generators designed to have the shaft spin half as fast, but if they want to be connected to the grid in North America, they are required to be 60 Hz generators, so they are built and designed to have four poles, two positive and two negative. Every time a generator shaft spins one revolution, it hits each of these four poles. It needs to be perfectly synchronized with the two-pole generators, so instead of the shaft rotating at 3600 rotations per minute (rpms), the shaft in this four-pole generator needs to spin 1800 rpm and will accomplish the same thing.

Take it a step further to a six-pole generator, with three positive poles and three negative poles on the rotor. In order for this six-pole generator to be in perfect synchronization, with the generators mentioned above, the shaft in this generator will spin at 1200 rpms. If they are all passing a positive and negative pole at the same time, they are essentially working together. This is called frequency. Frequency is measured in hertz (Hz), or cycles per second. Frequency is used as a measure of how well an electrical interconnection power grid is balanced between generation and load. That just might look and seem complicated and intimidating, but it's not. You might have gotten a bit bored and lazy. Go read it again. It's not that hard; it just sounds hard to understand if this is the first time you have heard about it. I'm going to keep explaining, so if you don't understand, just read the paragraph again.

When the generation supplying electricity to the grid is exactly equal to the amount of electricity being consumed by all of the loads on the grid, the frequency will be exactly 60 Hz. If the load of the interconnection is slightly higher than the amount of electricity being supplied by the generators, all of the generators will slow down equally as the different machines try to maintain speed (whether that's by making more steam or adding more gas or releasing more water, for example) to get them back to full speed due to the additional strain of trying to support the load. It is very similar to a portable generator that a carpenter might use to run equipment for a construction project at your house. Or better yet, maybe you have one at your house for use during emergencies. When the generator is running and additional load is put on it, you can hear a change in the pitch of the running engine as it slows down to "get more" fuel to bring it back up to normal speed. When this happens on the huge machine of the interconnection, the frequency is slightly less than 60 Hz, but when I say slightly, I mean very slightly. If the frequency falls outside of a certain threshold, power grid operators are made aware through audible and visual alarms in the control rooms they work in. So, going back to the earlier statement that the grid is the biggest machine in the world, let's put this into perspective: The thousands of generators are out there, all spinning around and synchronized to each other so they're all passing magnetic poles at the exact same time (related to the number of poles). If the load-to-generation balance is slightly mismatched in the interconnection, the generators all slow down at the exact same rate related to the number of poles they have. For example, if the balance of load to generation is 99.9% generation to feed 100% of load, then:

two-pole generators will spin 3600 X .999 = 3596.4 rpms;
four-pole generators will spin 1800 X .999 = 1798.2 rpms; and
six-pole generators will spin 1200 X .999 = 1198.8 rpms.

The frequency would be approximately 59.94 and low enough to cause every power grid operator's curiosity to peak, as visual or audible signals would be in every control room in the interconnection, alerting the operators that generation to load is mismatched considerably. The visual and audible signals would be the prompt to let the operators know they need to figure out if their individual areas are part of the reason the generators have slowed down, and if they discover that their area is causing the problem, they need to fix it immediately before the problem is compounded.

Do you recall the scenario of the power grid operator at South Illumini, LLC, in which the generator reduced 50 MW and the load of the area did not change? The reason the load did not change was because the tie-lines all shared an increase to allow 50 MW more to come into the system. But something did change; it was the frequency. The frequency slowed down slightly, too—maybe to 59.999 instead of 60.000, and it did this in the entire interconnection, not just in the South Illumini area. In this scenario, you, as the power grid operator of South Illumini, LLC, had a responsibility to the interconnection to increase 50 MW of generation somewhere else in your system as soon as possible to recover the frequency to 60.00. Guess what? Something else changed in the South Illumini area as well, and it's called your *area control error* (ACE) value.

The ACE Value

The ACE value is arguably the most monitored value in the control room by power grid operators. Well, what is it? As we explained, frequency is used to determine the balance of load to generation in the *entire* interconnection that you are a part of. ACE is used to determine how your individual area is balanced to its load and other obligations. Huh? Be patient; it is all going to be explained slowly. Do you remember when I showed you all of the tie-lines and summed them up into one number to make it easier for a power grid operator to assess? Well, the ACE value uses that principle as well. What does ACE mean? ACE indicates one thing: If you are doing your part to make sure the interconnection is balanced. One of the most important organizations in the industry is called balancing authority, or BA. Most RTOs or ISOs assume this BA responsibility for the huge area they are responsible for. BAs are sometimes referred to as balancing areas as well. These organizations are required to make sure their generation is balanced to their load plus any scheduled electricity they have bought or sold with their neighbors or elsewhere. The power grid operators, in the area they are responsible for, use the calculated ACE value to quickly determine what needs to be done. Okay, don't get scared, but here is an equation. The nice part about this equation is that you have already learned nearly all of the components. The ACE equation for individual Balancing Area's is this:

$$ACE = (NIA - NIS) - 10B\,(FA - FS) - I_{ME}$$

Where:

Net Interchange Actual (NIA) is the algebraic sum of actual flows on all tie-lines, which is the sum of the 18 tie-lines you had in South Illumini, LLC, the 400 MWs coming in.

Net Interchange Schedule (NIS) is the algebraic sum of *scheduled* flows on all tie-lines. Now, this one we have not explained, but it is not difficult. Let's assume you had an opportunity to buy 400 MW for the hour from your neighbors at a price cheaper than you could generate it. Economically, that is what you would do, and your schedule would become 400 MW of imports for the hour, which means NIS = -400. Notice the sign is negative. For the examples ahead, a negative sign is used to represent imports to your system and a positive sign to represent exports from your system.

So, the first half of the equation says you need to do your part to maintain your generation at a level equal to the load and schedules in your area. You measure what the actual flows on all of your tie-lines are right now and subtract the scheduled flows and then find the difference between the two, so now we have the left-hand side of the equation (NIA – NIS) complete.

$10B$ is the frequency bias setting (MW/0.1 Hz) for the BA. The constant factor 10 converts the frequency setting to MW/Hz. The larger the area (or greater proportion of customer demand) you are responsible for in proportion to the interconnection as a whole, the bigger the (B) bias setting is. I will explain this in more detail shortly.

Frequency Actual (FA) is the actual interconnection frequency—self-explanatory.

Frequency Schedule (FS) is the scheduled frequency. FS is normally 60.00 Hz.

Meter Error (I_{ME}) is the meter error correction factor typically estimated from the difference between the integrated hourly average of the net tie-line flows (NIA) and the hourly net interchange demand measurement (megawatt-hour). This value is normally very small or zero, so for the purposes of this book, you can dismiss it to make things a little easier.

This second half of the equation says you (as a Balancing Area) need to do your part to help support the interconnection system frequency. So, you measure what the actual frequency is right now and subtract the scheduled frequency (normally 60.00), and then find the difference between the two. That's the (FA – FS). You then multiply that number by 10 times your frequency bias obligation, which is the *B* in the equation definition above. The what? Okay, NERC requires BAs to have a frequency bias obligation to insert into the ACE calculation. It's based on the size of the balancing area compared to the overall size of the interconnection you operate in. It is designed to make sure that each area does a little bit automatically to help maintain the interconnection frequency within limits. It is intended to prevent the generators throughout the interconnection from slowing down or speeding up too much. If that occurs, protective relay systems start disconnecting equipment throughout the system to prevent permanent damage and could cause a cascading event, which you most certainly would read about in newspapers for weeks following it. Frequency bias obligations are numbers of MWs that power grid operators are required to have in their ACE equation per .1 Hz. The value has to be at least 1% of their forecasted annual peak load, so an area that has a forecasted peak for the year of 10,000 MWs would be mandated to have a bias setting of at least 100 MWs. Bias is always expressed as a negative number in the equation.

Most of this info is taken directly from the NERC BAL Standard.[17] The BAL Standard is only one set of rules that power grid operators need to follow, so it needs to be pretty straightforward. Right now, your eyes are glazing again and you're thinking about putting the book down, aren't you? Hang in there just a few more paragraphs and you'll actually understand how this works, with a few brief examples.

Let's use the South Illumini, LLC, example again as a reference point. You are the power grid operator of the area, and you want to know what your ACE is. Well, assuming you have scheduled 400 MW of imports to get into your system, you are probably going to calculate a very low ACE. The area you're responsible for is relatively small in comparison to the Eastern Interconnection, so we will assume your bias (*B*) value in the equation is 50 MW in the 10*B* portion of the equation:

$$\begin{aligned}
ACE &= (NIA - NIS) - 10B\,(FA - FS) - I_{ME} \\
ACE &= (-400MW - -400MW) - 10(-50)\,(60.00 - 60.00) - 0 \\
ACE &= 0 - (-500)\,(0) \\
ACE &= 0 - 0 \\
ACE &= 0
\end{aligned}$$

Your system is perfectly balanced in this scenario.

Now let's use the scenario of the generating unit that reduced 50 MW and the frequency declined to 59.99. We are also going to remove the IME value from the equation, as it does not have a significant effect on the results and/or the actions the operators would take.

$$\begin{aligned}
ACE &= (-450 - -400) - 10(-50)\,(59.99 - 60.00) \\
ACE &= (-50) - (-500)\,(-.01) \\
ACE &= -50 - 5 \\
ACE &= -55
\end{aligned}$$

This means your area needs to increase generation by 55 MW to get back to a zero ACE. A negative ACE value in a control area normally indicates the area is under-generating (or deficient), and a positive ACE value represents over-generating (supplying too much). That's it. If you understand this, you understand one of the most important numeric values that power grid operators are constantly monitoring and taking action to keep near 0 MW. This ACE value is calculated and updated at least every few seconds.

Let's do one more example of another area. Going back to your system in Illustration II-1, let's use one of the balancing areas neighboring your South Illumini area. The area neighboring you is scheduled to be selling 1000 MWs. This neighbor measures the actual flows on its tie-lines and sees that at this moment, it is sending out a total of 1050 MWs.

This neighbor has a 100 MW per .1 Hz bias (that means this neighbor serves about double the load you serve). The actual frequency is at 60.02 Hz, and the scheduled frequency is 60.00 Hz.

ACE = (NIA – NIS) – 10B (FA – FS)
ACE = (1050 – 1000) – (10)(-100) (60.02 – 60.00)
ACE = 50 – (-1000) (.02)
ACE = 50 – (-20)
ACE = 70

Since this company's ACE for this point in time is 70, they are required to try and correct it by decreasing generation in their area by 70 MWs. This is done by sending electronic signals to generators in their area to decrease output slightly. Remember, this is done, by requirement, at least every few seconds.

These are all just mathematical calculations to make the power grid operator's job of understanding multiple data points manageable. As discussed, some of the electric areas which perform these calculations are absolutely enormous. Doing the calculations for a very small area is reasonably manageable, but can you imagine the calculations for an enormous area that includes a dozen or more states with a couple thousand generators and several hundred transmission lines tying you and your neighbors together? The math is exactly the same, but there is just a lot more of it.

Constraints and Real-Time Contingency Analysis (RTCA)

Now, let's move on to the last building block, which is constraints. What are constraints? As you learned earlier, organizations that operate the bulk power grid are governed by standards mandated by NERC, listed in Chapter 6. One of the most important rules/standards of operating the grid is that it has to be able to withstand all credible N minus 1 (N – 1) contingencies. Uh, okay, what's that supposed to mean? N – 1 means normal minus (–) one single unexpected loss of equipment. The industry term for these unexpected occurrences is "trip."

Before we go too far, let's discuss transmission lines again briefly. Transmission lines can be compared to interstate highways because they can handle just so much traffic. When traffic on the highway is traveling at less than the posted limit, we call this congestion, or, in other words, a constraint. Electric transmission systems are no different. Earlier, we discussed various lines and voltage levels. These various lines can transport just so much electricity before they get congested. When they are fully loaded at 100% of their rating, it becomes an unfavorable, risky operating condition. There are many reasons for a limitation on a transmission line, which we will not go into now. The important part to know is that operators must ensure that no line will exceed its rating and also must prevent a line (or any piece of equipment) from exceeding its limit following the failure of another line or piece of equipment. This is done to ensure safe and reliable continuity of service to all customers.

This is how a power grid operator will see a potential problem: Go back to the South Illumini, LLC, area and put yourself back in the chair. Your computer monitor displays an alarm indicating that one of the lines from Hoosier Power Corp. to South Illumini, LLC, will go to 101.5% of its rating for the loss of one of the lines from Mizzu Energy–South Illumini, LLC. This 101.5% means the line will be 1.5% over the capacity the line was designed and built to carry ***if*** the other line comes out of service at that moment.

Remember that on the transmission system, everything is tied together, so normally, nobody's lights go out for a loss of equipment on the power grid (aka Reliability), so you will still be importing 400 MWs into your area even if you lose the Mizzu Energy–South Illumini, LLC, line. Everything is operating like it should, but what if any one thing happens? If the Mizzu Energy–South Illumini, LLC, transmission line unexpectedly goes out of service, the alarm tells us the Hoosier Power Corp.–South Illumini, LLC, line will overload, and that means it *could* also possibly go out of service on overload. Power grid operators take action to the alarm to prevent that potential overload from happening. In layman's terms, this means they need to plan and be ready for any single credible event that might happen on the grid. In this example, generation on either side of the problem will be instructed to increase or decrease, based on the positive effect, until the loading on the line reaches the desired level.

For an electricity market to operate as efficiently as possible, operators are expected to keep the major transmission paths between inexpensive generation and dense load areas loaded up near the limits of the lines to get the maximum amount of the lowest cost power possible to the high-load areas. Sounds like a good premise to save money, doesn't it? But here's the difficulty, as I explained just a bit ago: If any of the lines that are loaded up near their limit happen to unexpectedly go out of service for any reason (we'll talk about some possible reasons a little later), the remaining lines feeding into the load pocket are not supposed to go over their limits, either. This is again called protecting against the N – 1 contingency.

Now that you know this, let's try another example. Go back to the South Illumini, LLC, area and consider it is a different time of the year than in the previous illustrations. For this example, we will use a day in August when temperatures are in the high 80s. Take a look at your generation and tie-line flows:

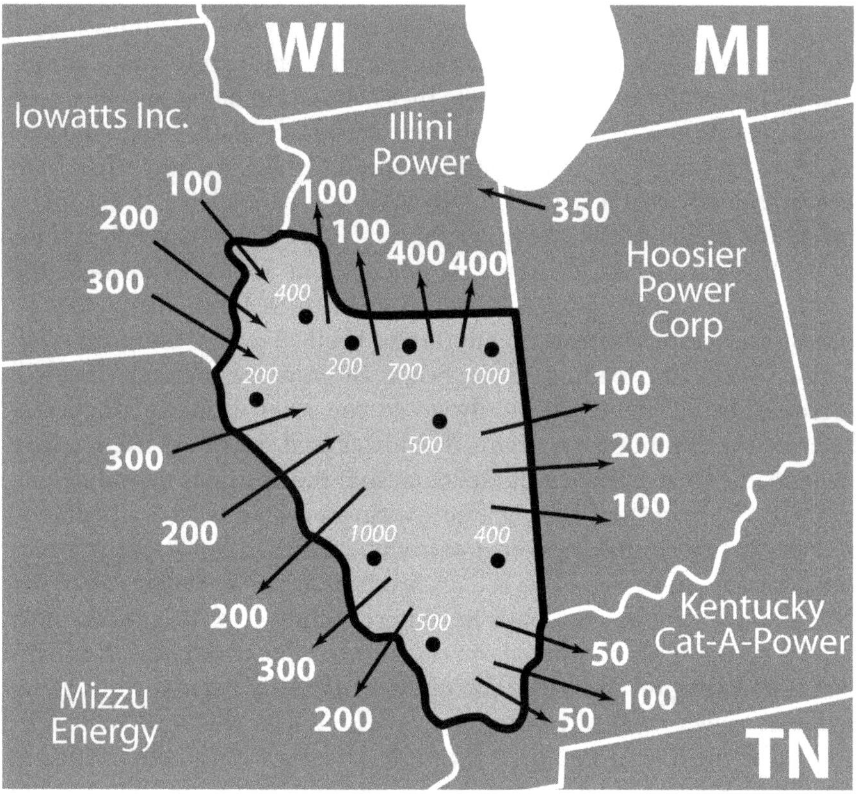

Illustration II-6

First, go ahead and calculate your load. Remember, it's your generation plus the net sum of your tie-lines. I will wait here for you to complete the calculations. I kind of feel like I hear the *Jeopardy* theme song playing . . . Bomp todododa . . . la la Bing . . . answer please? You should have come up with generation of 4900 MW and exports of 1200 MW, for a total load of 3700 MW in your area. Big difference from when you originally got the area. Today, the temperature in this fictitious company's area is an average of 88 degrees. Do you notice the tie-lines and the flows on them? It appears lots of MWs are heading to northern Illinois because of the high demand area of the city of Chicago and also that the tie-lines have heavy flow on them heading toward the city of St. Louis in Missouri. Last night, a great deal of generation was unexpectedly lost in northern Indiana and is not expected to return to service for a minimum of four days. It was not in your area, but it affects you because the loading on your lines will be higher than you would normally expect going to Indiana and northern Illinois with the unexpected loss of that generation. It is quite likely that with the loss of that generation, Hoosier Power Corp. is buying more power to replace the generation it lost in the evening.

So, you were notified of the generation outages by the Reliability Coordinator of the area. The RC is looking at the big picture and communicates important information like that generation being out of service to all operators who are affected. Your computer display is giving you an alarm telling you that South Illumini, LLC, tie-line with Illini Power, the one that is loaded at 400 MW, will be loaded at 108% of its rating for the loss of the Hoosier Power Corp.–Illini Power line. This is the line on the illustration that is loaded at 350 MW up near the south end of Lake Michigan. What happens if the line actually does go out of service before you do anything? Well, the first thing that will happen is the line loading on the South Illumini transmission line will increase. It will almost certainly go to the level of 108% predicted, because the computer programs the power grid operators use are very dependable to give them accurate data, so if the line does trip while you are sitting there, you have a line overloaded and must immediately reduce the excess loading on it extremely quickly. But there is a secondary concern as well. The RTCA program which calculates all of the N – 1 contingencies continues to run every minute or two, so the next time it runs, it puts another alarm on your computer monitor. Remember, the loading on the line is currently at 108%. What if any one thing happens now? Well, the computer is telling you this line will go to 136% for the loss of the line next to it, which was originally loaded at 400 MW. Power grid operators know it is unlikely the line will be able to handle that high loading for its rated capacity for any length of time so the line will almost certainly trip because it is overloaded if this second contingency occurs before you can get the loading down. This is what will likely begin a cascading event like you read about in the historical events section.

There are many more contingencies that power grid operators monitor and respond to. Take a look at this map of your area again, only this time, notice I zoomed into it (think of it as Google Earth) so you can see a lot more detail. Every single transmission line, generator, and piece of equipment inside the substations has a direct effect on the other parts in the diagram. Your South Illumini, LLC, computer calculates the what-if scenario for every single one of them, and it gives you the data to evaluate and take action for.

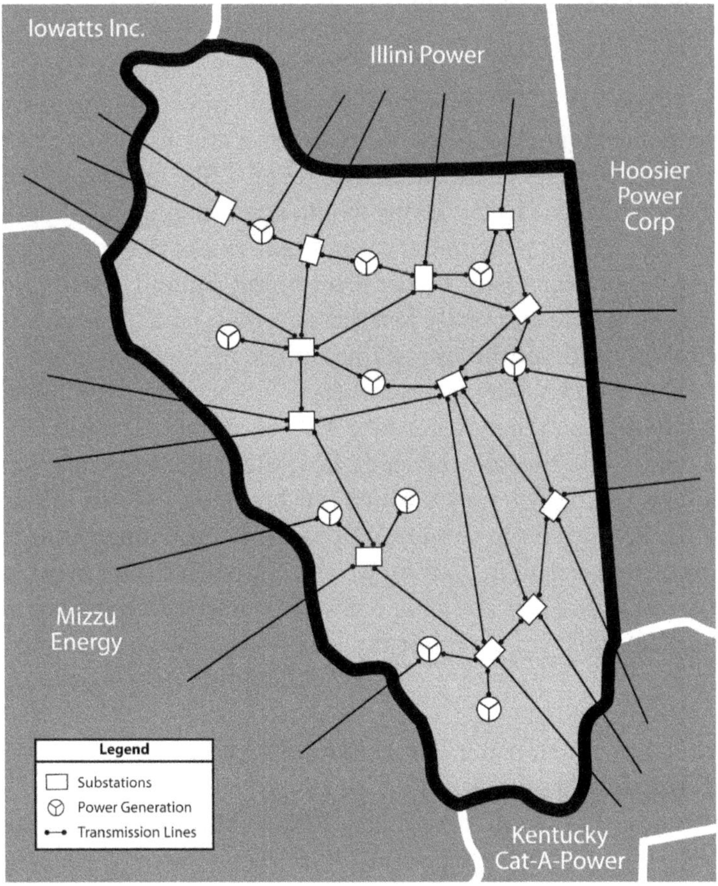

Illustration II-7

The RC for this area is also calculating the what-if scenarios for the loss of all of these bulk power grid elements and will work with the operators at South Illumini, LLC, to resolve any issues that develop.

Some Power Grid Contingencies That Cause Operators to Have Exciting Days

Generators and transmission lines and other electric grid equipment trip, which, as you learned, is an industry term that means they fail because of some type of mechanical failure or act of nature or any other unexpected reason. Soon, you will see some examples of things that have caused lines to trip, but first, a brief explanation of two terms: short circuit and overloaded circuit. We're sure you've heard both terms and even used them yourself. Here are two scenarios; you guess the cause.

Scenario 1: You had a 32" TV sitting on a stand in your living room; you have just gone out and bought a 60" plasma set and also got a new home theater with 200 watts of surround sound and a Blu-ray player. You also bought one of those electric strips to allow you to plug all of your new high-tech devices into the outlet in the wall, so you plug everything in. Now you're getting ready to fire everything up and watch the new super thriller movie and really test out your new stuff. You turn on the Blu-ray player and put the movie in; you fire up the home theater and set the volume up halfway and tell everyone to sit down and buckle up. Finally, you turn on the monster new TV set, and two seconds later, the Blu-ray, the home theater, the cable box, the lamp on your end table, and the light in the fish tank all turn off. Is this an overload or a short circuit? Yeah, that's right; this one's an overload. You plugged too many things in on one circuit in your house and tripped a circuit breaker. You reset the circuit breaker and plug a few of your new purchases into different wall sockets to spread out the load a little more, and everything works fine.

Scenario 2: You take a paper clip and unfold it to make a straight piece of wire. You then bend it in half to look like a tall U. You grab a hold of the bottom of the U and use it like a fork to put into each side of the wall socket. There's a brief instant of seeing a spark right before you shock the hell out of yourself and swear you will never try anything that stupid again. The circuit breaker in the panel box trips opening the circuit; most likely before you even finish your *stupid* thought. Is this an overload or a short circuit? Yeah, you know. Why is it called a short circuit? It is called a short circuit because you have created a shorter way for the electricity to complete its path than having to go through the circuit it was designed to flow through. You decide whether you're a genius or an idiot for doing it. I will politely keep my opinion to myself.

Some common scenarios out on the power grid that cause transmission lines to unexpectedly go out of service include an ice storm, snow storm, or even rain storm moving through an area. The ice begins to accumulate on the transmission lines, or a tree gets bent over from the weight of the snow, or a wet tree gets blown over from the wind. The conductor touches something that allows the electricity flowing through it to go to ground and causes a circuit breaker protecting the line from damage to open until the condition can be fixed. This is a short-circuit condition because the electricity has found a shorter way to complete the circuit to ground than the path it was built to flow through.

Utility companies spend millions of dollars every year patrolling and trimming trees around transmission lines, but sometimes things get missed or the trees grow faster than anticipated and one tree gets close enough to the line to let the electricity arc over to it, and then the electricity flows into the earth, causing a short circuit. The relays at the ends of the line open a circuit breaker to protect the line from melting or burning down. This is called a phase (or line)-to-ground fault because one of the lines that carry electricity has found a way to allow the electricity to make it into the earth. When this happens, the current flowing through the line will increase to incredibly high levels and will melt the wire it's flowing through if the line is not opened immediately. The relays protecting these lines from these types of faults will detect the super-high flows associated and open the circuit breakers at the ends of the line in a fraction of a second.

Most overload conditions are predicted and identified in advance by the RTCA program looking for N – 1 conditions, and power flows are reduced by power grid operators "re-dispatching" generators, which we briefly explained in the South Illumini area examples. I will elaborate further into how this is done in a little bit.

The power grid operators are monitoring what is happening on the grid, and they are prepared for all but the most freakish of combinations that might happen. Again, these are called N – 1 events or contingencies, as we discussed earlier. Power grid operators monitor everything that might happen; the next trick is to think; *What is the worst thing that could happen right now? What are the 10 worst things that might happen in my area of responsibility to the grid?* Once the operators think of these "bad" scenarios, the operators are creating mental plans for what their first steps will be to resolve that "worst" thing so they are prepared to react to the problem. Some power grid systems are looking at thousands of N – 1 contingency scenarios calculated by a computer every few minutes. Although there's not anybody smart enough to look at all the combinations of things that could happen every minute, the operators use computers to help.

Most large ISOs, RTOs, and RCs use incredibly highly advanced computer systems that collect real-time data from all of their important intersection points every few seconds. The data is then fed into a system called state estimator, or SE. The SE system then looks at all of the real data and compares the readings to the estimated data to make sure that they make sense. For instance, if you have 300 MW flowing into a substation but only 200 MW flowing back out, there might be either an error in one of the readings, or you might have an error in the estimating process. But which is it? Companies that have SE programs normally assign someone to look for mismatches 24 hours a day, 7 days a week. When a mismatch between actual flows and SE flows is discovered, this individual immediately begins investigating to find out why the discrepancy is there.

The next thing these highly advanced computer systems do is the Real Time Contingency Analysis to determine all of the N – 1 contingencies we have already described. The computers evaluate every possible thing that the engineers can find that might credibly happen. What that means is, you will have several electrical engineers sitting at desks every day and watching what happens, thinking about what could happen, and then building a contingency and adding it to a database so the computer will recognize it if it occurs sometime in the future.

Many SE applications run approximately every 60 seconds; as soon as SE "solves," which means that everything looks OK and the inputs and outputs look valid, RTCA will kick off and run the thousands of what-if (N – 1) scenarios. Each of the RTOs, ISOs, and/or Reliability Coordinator areas are made up of many different individual companies, and all of the N – 1 contingencies of each of those companies are monitored and discussed together by these organizations as needed. They are monitoring a wide area view and coordinate and communicate with all parties with an interest. This view is often referred to as the big picture. One of the many power grid operator positions is also called the Reliability Coordinator (RC). This is a specific individual operator in real-time operations. These individuals are responsible for monitoring all of these N – 1 conditions and ensuring there are plans in place to reliably mitigate or correct any of these anticipated conditions. We will get into this in more detail in the following sections, but suffice it to say the Reliability Coordinator as a company, and an individual with the title RC, is ultimately responsible for ensuring these scenarios are all able to be resolved. As mentioned previously, the RTOs and ISOs accept the Reliability Coordinator responsibilities in their organizations and have staff with the RC job title.

If generator A trips, what will the flows on all of the lines feeding into the area go to? Or in another case, what if transmission line XYZ trips? What level will other lines in the surrounding area go to? Will any of them overload? Power grid operators want to know what all of the projected flows will be for all of the N – 1 scenarios.

Take a single transmission line and multiply it by the number of major generators and transmission lines in the area. The reason this is done is because it's all connected and each piece of equipment has an effect on the other when they go out of service. Some pieces have a big effect while others have virtually no effect, but that mathematical number of evaluated possible situations adds up pretty quickly. RTOs, ISOs, and Reliability Coordinators monitor thousands of these types of scenarios. Every few minutes, the scenarios are calculated by the RTCA computer program, and the power grid operators get the results again and again. Additionally, every time another scenario is found that could realistically occur, it goes to that same number + 1 more. The new contingency gets added to the database, and the results are calculated, along with all of the others. If the computers start slowing down because they are looking at too many possibilities, the organization adds more computers. There are never too many possibilities to evaluate. Today's advanced computers have allowed power grid operators to have a lot more visibility into what might happen than they were able to see 20 or even 10 years ago, because of how much computers and the industry as a whole have evolved. The what-if scenario gives power grid operators the additional time needed to evaluate and plan corrective action just in case the event occurs.

Electric Wholesale Markets

As mentioned a bit earlier, the large RTOs and ISOs run a wholesale electricity market in addition to performing the Reliability Coordination services. Part of that means they evaluate all the offers they receive from the generators across their entire area. Some markets are defined by state boundaries, and other markets operate across all or part of numerous states. The RTOs and ISOs then evaluate how much electricity is forecasted to be used to meet customers' demand tomorrow. In most electricity markets today, there are actually two separate markets: the day-ahead market (DAM) and the real-time market.

The founding premise of belonging to an RTO or ISO is that if a bunch of electric utilities combine and then use one central market clearing process for the combined and much larger geographical area, they will come up with an overall more efficient base set of generators than if each utility were to try to cover its own needs. In order to actually realize these efficiencies, though, in this type of environment, they will be trying to move as much electricity as possible from the areas of the market where the cheap generation is available into the metropolitan areas. Generation is generally not as abundant in metropolitan areas due to a multitude of reasons including, but not limited to, lack of available land for generating site locations and stricter environmental restrictions (noise/pollution/coal dust), but metropolitan areas always represent a large percentage of the overall demand of a market area. So if that is the generation commitment design, then the transmission lines feeding the high-load areas will often be maximized to their operating potential.

Security Constrained Unit Commitment (SCUC) and Security Constrained Economic Dispatch (SCED)

SCUC and SCED are the foundation of what allows power grid operators to do their job in large RTO and ISO market structures. Security

Constrained Unit Commitment (SCUC)[18,19] does exactly what its name says, and a little more. SCUC is a software program that organizations use to commit generators based on their offers on a day-ahead basis to make sure enough generation is on and can reliably feed the load in the most cost-effective manner. Security Constrained Economic Dispatch (SCED)[20,21] is the software program used in real-time operations to fine-tune and adjust the SCUC commitment as needed.

SCUC

This is going to be extremely simplified, but I will provide an explanation to illustrate what an SCUC software program does. The piece being discussed here is how a SCUC program works in a DAM. One of the many power grid operators sitting at a console with several computer monitors in front of them logs onto their console. They have a specific time that their market rules allow them to close their market bidding period for tomorrow. Let's use 9:00 a.m. for this illustration. At 9:00 a.m. exactly, their software automatically shuts off the ability for the member participants in their market to update any more bids or offers into the system. The operator then pushes a couple of buttons and enters some commands into their computer, and the computer begins collecting all of the data that has been submitted by the generator owners of each individual generator. Here's *some* economic and physical data the system software looks for and collects regarding generators for tomorrow:

1) Cost to start the generator *(This is not the cost to run it; it is just the cost to get the generator warmed up and ready to produce electricity.)*
2) Lowest output the generator can run at reliably *(also known as minimum output)*
3) Highest output the generator can run at reliably *(also known as maximum output)*
4) Cost to keep the generator running at minimum level per hour ($ per MWH x # of minimum output)
5) Cost to run the generator at every level between minimum and maximum level of output (This is known as a cost curve.)
6) Shortest amount of time the generator can run before it can come off (minimum run time)

7) Longest amount of time the generator can run before it has to come off (maximum run time)

Pretty simple, isn't it? Not really. How much does it cost to start the generator? Generator owners need to figure out how much fuel it will take to get the gigantic boiler sending steam to the generator warmed up and ready to go. Then they evaluate how many additional people are needed to be at the generator if it is actually going to be running. Then they add in the annual maintenance costs and taxes they will need to pay for the property the generator is on, and many more things I have not thought of, and figure all that into the economic offer or bid.

Then a separate process includes the companies that need to supply electricity to the customers in their service territory. These companies will bid in how much they are willing to pay for electricity in the DAM. This is a strategic move on the companies' parts. Ordinarily, the DAM is more stable and by market design should normally be lower cost than the real-time market (I plan to get back to the why this is true in a bit, but for now, assume it's true.), so the company that actually sends you your power bill forecasts that tomorrow, they're going to need 5000 MW of electricity from midnight until 1:00 a.m., then they're going to need 4800 MW between 1:00 a.m. and 2:00 a.m. because more people will be in bed and it is predictable behavior by many people doing the same thing night after night. For example, many homes have programmable thermostats that automatically turn down the heat a few degrees when it's bedtime. Electricity usage is a pretty standard curve, as we talked about earlier, so companies have a pretty good idea of how much electricity they're going to need tomorrow.

The companies supplying electricity to the end-use customers also have a pretty good idea of how much electricity is going to sell for in the DAM because they've been doing this for a while, so if they think the price for electricity in tomorrow's DAM between midnight and 1:00 a.m. is going to be about $30 and they're going to need 5000 MW for the hour, they might bid like this:

Hour Ending 01:00 $30/4700 MWH
Hour Ending 01:00 $29/100 MWH
Hour Ending 01:00 $28/100 MWH

At this point if any of the member companies serving their customers has said it's willing to pay $30 per MW for the first 4700 MW of electricity because it is betting the price for electricity tomorrow is going to be higher than that, but if it's cheap enough, they will buy another 100 MW and take the risk that the price tomorrow won't be lower than $29, and if the price in the DAM is lower than $28, they'll even take the risk that they won't lose money at $28; most importantly, they are betting that they have estimated accurately how much electricity they will need tomorrow. Below is a graph to help illustrate what the load trend traditionally looks like for many different balancing authorities. If weather conditions tomorrow are expected to be similar to today's, they will be able to predict their needs fairly closely. The left-hand portion of MWs would be scaled to the historical data archived by these organizations.

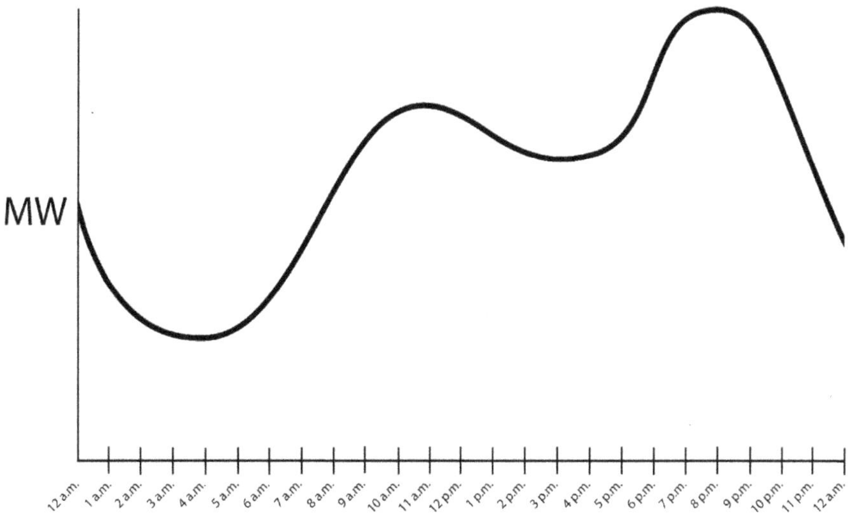

Typical 24 hour load curve pattern Illustration II-8

The SCUC software program commits enough generation starting with the lowest cost generators and then the next cheapest and so on until it has enough generation committed to meet the combined bid load of the entities described above. It then runs the second part of its software package to see if the generators that were selected in the commitment will overload any of the lines under any of the N – 1 scenarios we talked about. If the program predicts an N – 1 problem in the Day

Ahead Market, the software will commit generation in a location to help relieve the overload(s) and re-runs over and over until it gets the lowest cost overall solution to meet the estimated bid load while making sure all of the N – 1 scenarios can be handled reliably tomorrow. It will keep putting on the lowest cost generators in a specific area and taking off the highest cost generators somewhere else on the system (if possible) until it can mathematically resolve the reliability N – 1 constraints it detected. That's why it's called *Security Constrained* Unit Commitment. Once the final solution is complete, a power grid operator reviews the results to make sure all of the N – 1 constraints are able to be maintained below their limits and there is enough generation committed to meet the forecasted load, and publishes the results. These results are then sent to all of the market participants, informing them which of their generators have been accepted to run tomorrow. It also informs each of the other (load serving) companies how much of their forecasted load was covered in the DAM to allow the companies to see how much they will need to cover in the real-time market. Finally, these results include the locational market price (LMP) for each substation and generator in the market. These LMPs represent how much the loads connected to the substations will have to pay to cover the electricity being fed from each substation and how much each generator will be paid for the electricity produced at its specific location. More on LMPs coming up.

SCED

Okay, here we go. This is one of the most important pieces of real-time operations. If you think you've had enough, take a break and come back to this later. SCED is a software package that in real-time gathers data from across the entire organization's system and pulls the data in as an initial system condition or look at "where am I right now." It is completely separate from the SCUC program you just read about. The only way the two programs are related is that the SCED is using the SCUC generation commitment results from when it calculated them "yesterday".

Most SCED packages today run every five minutes (a new case every five minutes) on the cardinal five-minute marks with a 10-minute look ahead. What this means is that at 9:00 a.m., it will kick off a new case and the results will be sent electronically to the generators in the market entity's service area at 9:05, informing the generators where they are expected to be at 9:10. At 9:05, a new case kicks off that is sent electronically at 9:10 and notifies the generators where they are expected to be at 9:15, and so on. This happens 288 times each and every day. For an example, we will use the case that kicks off at 9:05 a.m.

At 9:05, SCED gathers a snapshot of where all of the generators in the balancing area's system (which remember *is* the RTOs or ISOs) are loaded at right now. We will say for illustrative purposes that, all together, the generation is loaded at 10,000 MW and is exactly where an earlier SCED run told each individual generator to be. SCED then looks at where the NIS (Net Interchange Schedule, which, as you hopefully remember, is a value in the ACE equation and is the sum of all of the energy transactions the area has made) is at right now and where it is scheduled to be at 9:15. For this example, we will say the NIS at 9:05 is importing 500 MW. Based on these numbers, we know that the expected load for 9:05 was 10,500 MW, because the generation plus the imports equal 10,500 MW. SCED then takes the NIS for 9:15 and sees that the NIS is scheduled to be 400 MW importing, so the balancing area is scheduled to lose 100 MW of imports. If the load is expected to be the same in the next case, then the 100 MW of imports will need to be made up for with additional generation in order to maintain load to generation balance.

SCED analyzes all of the offers for the generators that are online and how fast each of them can increase generation (ramp), and it solves for the lowest cost generation to pick up the 100 MW needed in the next five-minute case. So for this example, let's say the lowest cost generator that is currently online can ramp up at 5 MW per minute, is currently loaded at 300 MW, and has a maximum output of 500 MW. We know it can get to 350 MW in 10 minutes; therefore, at 9:10, when SCED sends the instructions out the door, it tells the generator to go to 350 MW. It then finds the next lowest cost generators and continues to do this same process across all of the offered generation until the 100 MW increase is met. End of sentence, paragraph and story? No, this is just half of the SCED process; this is just the economic dispatch piece. There are a lot of things that are not mentioned in this brief example for the sake of simplicity. For example, what if the generator in the above example had been already loaded to 475 MW? SCED would have only been able to raise the generator 25 MW instead of 50 MW because there was only 25 MW available on the generator before it reached its maximum output level. What if the generator had been instructed to move down 25 MW in the SCED case that had been sent out the door at 9:05 for 9:10? SCED would then need to limit the expected output for 9:15 based on the previous cases' instructions. There are a lot of moving targets in the SCED process, but for illustration purposes, it was kept as simple as possible.

Moving on to the security-constrained part of SCED, this is where the real complexity begins. Again, as mentioned numerous times, power grid operators are monitoring all of the N – 1 contingencies in their areas. Again, an area can be one state or, in some cases, all or part of numerous states. If an operator sees an N – 1 scenario that would overload any element in their area, they have some options on how to make sure the contingency doesn't result in extreme overloads or even possibly cascading outages.

One of the first options for any balancing area, large or small, is for the reliability coordinator to tell the organization with the issue to move generation out of economic cost order on the constraint that is approaching conditions where the N – 1 outage would cause an overload. This off-cost method is known by several different terms throughout the industry. Some regions call it OOM (out of merit); others refer to it as solving the constraint; and some define it as binding the constraint. Essentially, it means the power grid operator changes a parameter in their SCED software to make sure the line or equipment approaching a limit isn't violated. The software accomplishes this by moving generation in an "uneconomic" fashion. This is called OOM order dispatch, or off-economic dispatch. It isn't as bad as it sounds and is actually the right thing to do, as you will see below. If SCED just moved all generation in economic order only, eventually, lines would just overload and trip out of service because they would reach an overload limit. As mentioned earlier, most metropolitan areas do not have sufficient cheap generation within the city boundaries to feed their entire load. It's simply a matter of the availability of land and ability to site all of the needed generation within big-city limits. For these big cities, it is more cost effective to import electricity from surrounding rural areas where generation is more abundant than load.

Electric markets, as well as smaller areas that operate outside of markets, have a basic principle they need to abide by to serve their customers: meet (supply) the load reliably at the lowest cost possible. So whether someone is a power grid operator looking over a large big-picture area or a power grid operator watching their own smaller defined area similar to the South Illumini area you were working in earlier, they need to move as much inexpensive electricity as possible from the outlying areas into the metropolitan areas to displace more expensive generation in the more populated areas, but it has to be accomplished without overloading the lines. When power grid operators notice an N – 1 condition that could impact reliability, they make sure the SCED software moves generation out of cost order to solve for the condition. This is done by the SCED looking at all of the generation available to move and coming up with generators to raise that will help, and generators to lower that will help, and then using this subset of all of the online generation to come up with the lowest cost way of solving the constraint. If the software program moves a generating unit up 10 MW to help reduce the loading on a constraint, it also needs to find 10 MW of generation to move down in order to keep the generation-to-load balance.

The NYISO is one example of a semi-large power grid organization that combines the balancing area, reliability coordination services, and market functions together and handles these responsibilities inside the New York state borders. New York is a common example used throughout the industry to explain how market prices react to a constrained grid because of how understandable the geography is. Cheap power is predominantly in upstate New York, and high demand is in the downstate New York City and Long Island area. Explaining it is easy; being a power grid operator responsible for it is another story. The transmission system in New York is one of the most constrained systems in the Eastern Interconnection. It would be fun to tell you what an easy job the New York power grid operators have, just to ruffle their feathers and get their blood pressure up, causing them to give me the one-fingered salute they are so famous for, but the truth is, the power grid operators throughout the entire state of New York are very good at operating the state's electric system.

Illustration II-9 is an example of what a 24-hour period of five-minute SCED LMPs might look like if New York had only two pricing zones.

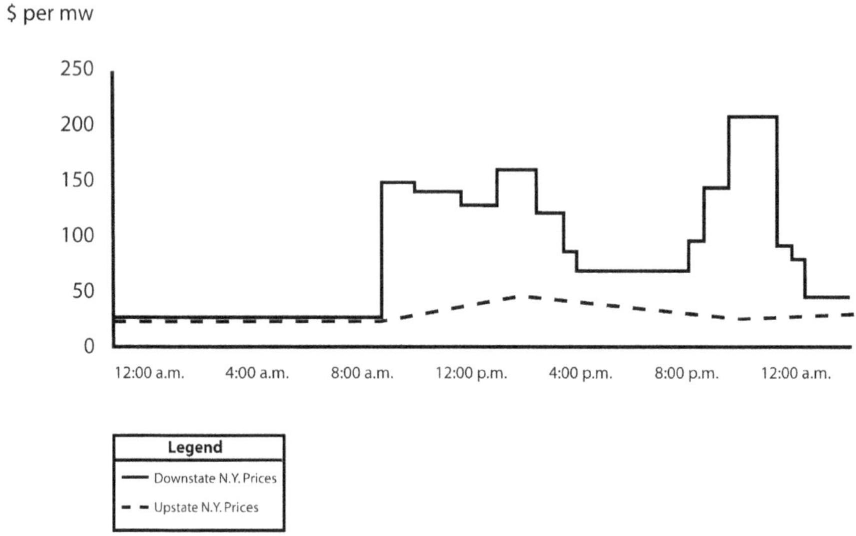

Illustration II-9

In the mock pricing chart, as you can see, prices change significantly in the New York City and Long Island areas as a result of a constrained system. The price spikes upward when the system is constrained as the load increases and more generation is needed. When the system is not constrained, the prices remain equal to those in the rest of the state.

So, SCED is run every five minutes and sends out these dispatch instructions to the generators where they need to move to meet the load in a reliable, cost-effective manner. This is constraining the generation to meet the security of the overall system in an economic dispatch. Note that I managed to put in all of the words for Security Constrained Economic Dispatch. Go ahead and read the paragraph again . . . you've got to love it.

The SCED not only tells each generator what level to generate at; it also generates an LMP for many locations on the system. LMP indicates the cost that the generator receives for the output it is instructed to move to. This LMP gives the power grid operators an indication that they are solving a transmission constraint. If they see that at one end of a constrained transmission line, the LMP is $50 and at the other end of the line, the LMP is $12, they know they the software is providing price incentives to generators at each end of the line. The software is sending the higher prices to generator that will help the constraint if they increase output and lower prices to generators that will help the constraint if they lower their output. The LMP is what every generator at that particular location will be paid for the power it injects into the system. It is also the price that every wholesale customer taking power off of the system at that location will need to pay for it—it is the marginal cost, or cost to supply or consume one MW of power at that particular location.

This is the very root of what a market is designed for. If the LMP at a particular location is almost always higher than the LMPs of the surrounding areas, it becomes an indicator that this particular location would be a great spot to build a generator, because the value of MWs is high in this area. If the LMP is almost always low at a particular spot, it is an indicator that there isn't any more generation needed at that spot, deterring companies from building in that spot. But again, here's the problem: The LMP is almost always highest inside of a large metropolitan area, where it's almost impossible to build a generator. The LMP is also almost always lowest in the middle of big rural areas, the same areas where no additional generation is needed but are the easiest to build generation at. So think about it for a few seconds and you'll see the logic of a market. Where is electricity most valuable? Is it in the middle of Chicago or in the farm fields of Iowa? It is basic supply and demand. The differences in LMPs across a market don't just tell you where it's best to build generation; they *also* give companies the incentive to build transmission. Companies realize that if they build transmission lines from cheaper areas to more expensive areas, they can transport the power across their lines and make a profit selling electricity into the more expensive areas.

Again, this is the very premise that markets were designed to promote: Show where electricity is most valuable and give companies the incentive to build generation plants and/or transmission lines where they are most needed. As a final note on LMP, if there is no congestion across a market footprint, the LMP will be the same across the entire area. This just means that if one more MW is needed anywhere on the entire system, it can come off of the lowest cost resource available at that particular moment.

SCED also does some other things that we'll touch on briefly and get into more detail in a few minutes. A balancing authority is required to carry regulating reserves and contingency reserves to meet normal and abnormal changes that can happen on its system. These reserves combined are called operating reserves, and SCED assigns these reserves to specific resources based on cost (this is a separate market called an ancillary services market). Different SCED packages assign these reserves in different ways, but all have one thing in common: They reserve capacity on specific generators or other resources to be able to quickly provide additional electrical output to meet conditions not accounted for in the five-minute dispatch in the SCED process.

Regulating Reserves

Balancing areas are required by NERC standards as well as by the "good utility practice"[22] that governs the industry's operation to maintain some additional generation that is online and connected to the grid that is able to respond to automatic generation control (AGC)[23] corrective signals continuously. These "reserves" are MWs that are earmarked on specific generators to respond to the regulating signals sent from AGC to correct the ACE explained earlier.

In SCED, the signals for the generators to respond to are sent every five minutes, so the resources are expected to be able to see where the SCED wants them five minutes from now and can move there in a straight-line fashion. Regulating signals are different. These signals are sent every few seconds, and there is a good reason for it.

Regulating generation resources, which receive the regulating signal instructions, are required to be on AGC and to be able to respond to these signals sent from AGC. Since SCED is run every five minutes, it is expected to be pretty accurate as to what the load will be, what the net interchange schedules will be, and where the generation is expected to be at the end of the five-minute SCED cycle; however, during the period the five-minute cycle is developing and the generators are ramping to their new output, overall, it is never 100% correct. Never! Think about it. If a market area has an expected load of 55,045 MW and a scheduled NIS of 1500 MW importing, the SCED software will tell the generators in its area to move to 53,545 MW. Over the next five minutes, we will assume all of the generators move at exactly the rate they said they could in their offers. We will also assume that there are no changes in the net interchange schedules due to extenuating circumstances. We will even assume that the forecasted load for the end of the five-minute period is exactly right to the last MW. What, then, could possibly be wrong? Well, at the end of the five-minute period, the math will be exactly right, but for four minutes and 59 seconds, or at least most of those seconds, it will be off at least a little bit because of you and me. That's right: We're the problem! Sounds a bit sarcastic, but here's the truth. People don't time it so that over the five-minute period being discussed here we all coordinate so we turn on our lights or heaters or toasters in a methodical fashion across the market area. We turn on our lights when we feel like it; we turn on our ovens when we're ready to cook. So, the load grows across the area in a spiked up-and-down pattern, not in a straight linear pattern.

It's not a big deal; it's just how it is. It's expected and is one of the reasons organizations have regulating reserves. Another reason is that it is known at the end of the five-minute period other factors will much more often than not have been at least a little bit off, such as the load forecast or the generators will not all be at exactly the level that they were requested to be at. Remember, we're talking about hundreds or thousands of generators all trying to make sure they're feeding exactly enough fuel into their system to achieve exactly the output the software requested.

The resources that are carrying regulating reserves make up for these "expected" errors by responding to AGC signals that are sent out each time ACE is calculated. Assume that the economic signals sent to the generators from SCED are expected to be followed in a straight-line fashion over the five-minute period. AGC then sends signals every couple of seconds to the regulating resources to deviate from that straight line to help correct the market area's ACE and make up for any differences in either the assumptions used to calculate the initial SCED results or deviations in the output of the generation resources.

So basically, regulating reserves are required to make up for "normal" or expected operating deviations to maintain an ACE close to 0. This ensures that load and the sales or purchases the area made are balanced to generation as close as they can be.

Contingency Reserves and Disturbance Control Standard (DCS) Events

So, if regulating reserves are required to make up for normal deviations, why are contingency reserves needed? You guessed it—to make up for unexpected or abnormal deviations. Contingency reserves are different than regulating reserves in a couple of ways, but the main difference is that they are not controlled directly by AGC action. In general, contingency reserves are only activated or deployed by the manual action of a power grid operator and are only to be used when necessary to balance load and generation due to an unforeseen event within the area. NERC standards also require these areas to hold back capacity on the resources in their areas to make sure that if these unusual events occur; rebalance can be achieved quickly without becoming a burden on the interconnection. Operating any longer than necessary to correct the imbalance is absolutely frowned upon at all levels of the industry.

If a balancing area unexpectedly loses a large generator, the frequency will drop in proportion to the size of the generator compared to the size of the interconnection. For example, if a 1000 MW generator trips in the Texas Interconnection, the frequency will drop more there than if a 1000 MW generator trips in the Eastern Interconnection. The frequency in either scenario will remain low until someone picks up the amount of generation that was lost; therefore, the area with the loss is imposing a burden on the interconnection by dragging the frequency down. The area is required to correct this imbalance in defined timeframes depending on the size of the loss. Large losses are required to respond based on another requirement in the NERC BAL Standard called Disturbance Control Standard,[24] or DCS. In the most basic terms, the DCS is a level that is no more than 80% of the most severe single contingency (MSSC) within the BA. Each year, all BAs are required to review all of the possible contingency losses in their areas and to define the single worst thing that might happen to cause a loss of electrical supply within their areas. Normally, this is the largest generator in the area, so as an example, we will say the BA has a large nuclear generator that, if tripped, will result in a loss of 1200 MW of electricity to the grid. This generator would then be identified as the BA's MSSC. Eighty percent (80%) of 1200 MW is 960 MW, so for this BA example, any loss greater than 960 MW will be classified as a DCS-level event. If any generator within the area is loaded at greater than

960 MW trips for any reason, the balancing area will be required to prove to its respective regional entity and to NERC that it recovered its ACE back to the lesser of 0 or the *pre-disturbance* ACE within 15 minutes from the time the generator tripped. This is to ensure the BA did not rely on or drag on the interconnection for an extended period

Clarifying note from above: Why the lesser of 0 or pre-disturbance ACE? If a BA happened to be over-generating by a substantial amount and was trying to reduce its ACE to get its system back in balance just prior to the loss of the resource, it doesn't make sense to require the area to go back to an over-generation condition, so it is only required to get back to the 0 ACE. If the BA was under-generating due to other normal conditions just prior to the loss of the generating resource but was within normal control conditions, it is only required to make up at least the amount of the loss in the first 15 minutes; then it can resume normal operation to recover its ACE back toward 0. It is fair without being overly burdensome.

After the loss of the generating resource and after the contingency reserves are deployed comes the next condition that is important: The area is required to *restore* all of its contingency reserves within 105 minutes from the time of the initial loss. This may seem like a long time, but again, this is designed to allow the BA to start additional generation or to arrange for purchases from its neighbors in a timely manner. This time period is again designed to make sure the BAs can reliably meet their load-to-generation balances in the most efficient manner. You need to remember, as I have attempted to express clearly throughout this book, that the overall concept of running the power grid is to always make sure the reliability of the grid is ensured as economically and efficiently as possible for everyone.

The next section of the book is a transition back to recreational reading which offers some further interesting things to read about, such as Japan's electric grid situation, the Smart-Grid future, the environment, the cost to run some of your household appliances, and other topics. I fully realize for the non-electrical grid person how much information you just went through, and how challenging it was. I hope you enjoyed the book, and if you did please spread the word through your favorite social media sites, such as LinkedIn or Facebook, and let others know the book is available. Thank you for taking the time to learn more about Power Grids.

PART III

Interesting Facts and Additional Information.

Electric Grid Troubles in Japan

EARLIER, YOU LEARNED THAT NORTH America operates as four separate machines (interconnections). Now you will learn about the amazing difference between North America's power grid and Japan's power grid. Japan is a country that is just a bit smaller in square miles than the State of California. California, population 30 million, is part of the gigantic interconnection called the Western Interconnection. Japan, in the smaller geographical area, has a population of nearly 130 million and is not part of any major interconnection, due to it being an island. Keep this in mind when you start to learn about Japan's power grid.

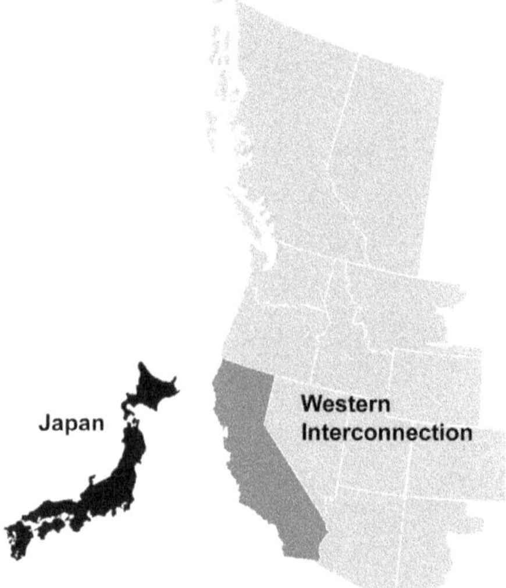

Illustration III-1

The horrible tsunami on March 11, 2011, that devastated much of the northeast coast of Japan created electrical power grid issues for the country. Several generators were destroyed by the tsunami. The most well-known, through media coverage was the three generators at the Fukushima Daiichi Nuclear Power Station. The reported radiation leakage is obviously a huge concern but not something I, or my solicited helpers, have enough knowledge to write about. What I am writing about is the loss of the generating capacity this created for the country. Under normal circumstances anywhere in the world, the extended loss of huge generating capability is something significant that area industry leaders would have concern about. The loss of these generation resources in Japan is exponentially worse than in most areas of the world.

Surprisingly, Japan operates two separate power grids. The system looks similar to that in Illustration III-2 diagram below. This diagram gives a feel for the heart of Japan's current power grid dilemma.

Japans two separate power grids   Illustration III-2

The following explanation isn't from exhaustive historical research but rather from simple research and some intuition on my part. When Japan began electrifying the country more than a century ago, the local city utilities were each operating very small local grids that weren't interconnected. The large cities in the northeastern part of the country, like Tokyo, purchased their generators from German manufacturers that operated on a 50 Hz system. The southwestern part of the country, however, purchased their generators from US manufacturers that operated on a 60 Hz system. When the country began to interconnect small grids, it ended up with two separate interconnections. You have to remember that in that time period, (early 1900s), people weren't traveling hundreds of miles regularly and most of the small towns had no electricity. The Japanese government probably didn't realize at the time what each town used for its own electricity because people weren't carrying hair dryers and laptop computers with them from one town to the next. As long as the lights in the big hotels came on, everyone was thrilled. The political leaders may not have even known the country was dividing itself into two different power grids until it was too late.

If the Japanese got a chance at a few historical "can we do that over" opportunities (golfers call them mulligans), then the design of their power grid would probably be one opportunity they would choose. The two interconnections are connected by a few converter stations that can convert the power to the other interconnection's frequency, but only to a *very* limited level. Since the tsunami in 2011 knocked a lot of generation out of service in the northeastern part of the country, the northeastern interconnection has had to institute rolling blackouts due to insufficient capacity while the southwestern interconnection has excess power but can only transfer a limited amount due to limited system transfer capability through those few relatively small converting stations.

The following excerpt explains the situation:

> *Before the quake, Japan's total generating capacity is about 230,000 MW. These capacities are spread between the west grid (peak load 98,000 MW) operating at 60 Hz where major cities like Osaka, Kobe are located, and the east grid (peak load 82,000 MW) operating at 50 Hz supplying power to Tokyo and the quake damaged region of the Northeast. Up to 21,000 MW of capacities were not available after the quake in the east grid. Even with partial generation restoration, aggressive conservation efforts and emergency generators, there is still a power shortage, some esti-*

mated as high as 9,000 MW, to meet the summer peak load in the east grid.

Although the undamaged west grid has spare capacity, its transfer capability to the east grid is restricted by the differences in frequencies. The 50/60 hz differences require a back-to-back ac-dc-ac conversion and at present there are three converter stations at the boundary of the two grids with a capacity of 1,200 MW. Expanding that transfer capacity is already under discussion by government and utility officials as to how quickly such converter stations can be added to boost transfer. Cost estimate for an addition of 1000 MW transfer capacity would be in the $800 million range.

One of the lessons learned from the quake is the reliance of electricity to keep Japan's 59 nuclear reactors in a safe state. Even with backup generators, grid power is essential to provide long term supplies. Therefore, the new grid must be able to withstand disasters with adequate redundancies and be able to be restored quickly. Smart grid technologies which can pin-point faults and self-heal would greatly enhance grid reliability.

Before the quake, Japan had called for a 50% increase in nuclear power for its base load generation. While the future of nuclear in Japan is not clear, the increase reliance of renewable, such as wind and solar in Japan's generation mix, looks more likely. A flexible grid to deal with these intermittent power sources as well as base load nuclear is essential to ensure reliability, in particular under the intensive air-condition load summer period. These would call for increased use of high power electronics to regulate power flow and dc technologies to pipe power from off-shore wind to coastal load centers.[25]

A 50 Hz system and a 60 Hz system are completely different and can't be tied together except with expensive conversion equipment that takes years to build and install. Japan does have this technology, but as you read above, it is limited to approximately 1000 MWs. This leaves the country with a problem due to the generation supply shortage in the northeast grid. Demand has often been greater than supply in the northeast grid, and as a result, Japanese power grid operators have had to institute rolling blackouts. Power grid operators in Japan have a very difficult job, and it seems like it's going to be this way for a while. Japan has hired several generation-construction companies to build new generation, but this takes time to plan and build. The normal schedule for planning and building a generating station is normally multiple years; however, under the dire situation the country is in, efforts have been made to severely reduce the normal length of time.

Some people traveling from North America to other parts of the world may read about the above situation in Japan and be a bit confused because of their experiences traveling to Europe. Much of Europe operates on a 50 Hz system. North America and many other areas in the world operate on 60 Hz systems, but when the worldly North American traveler takes a laptop on a business trip to Great Britain, the charger can plug into the wall in the hotel room and there are no issues. This is because technology has been created for the laptop or cell phone *charger* to recognize what type of power grid it was just plugged into and to automatically adjust the charging of the battery without harming the device. But if you're going to move to Europe for a new job, make sure you negotiate your salary to take into account the fact that you'll need to buy a lot of new appliances. You can fly your crock-pot or vacuum cleaner over with you, but it won't work when you plug it in unless you buy some very expensive converter equipment.

Smart-Grid Technology and What It Means to You

The term "Smart Grid" has as many meanings as there are ants in and around an anthill. Just do an internet search for "Smart Grid" or "Smart Grid technology" and see for yourself. Overall, the term basically means more automated pieces of equipment are being added to the electric power grid. But "at what level?" is really the big question that will be answered over the coming years. The grid right now is actually already extremely smart. It has protective relays to take pieces out of service to avoid damage, and it has meters on pretty much every transmission and distribution line to feed into computer systems and calculate if everything is functioning within limits. It is able to be operated all day every day with hardly ever a hitch.

Now, many electric utilities across the country have a "time of use" electricity program. The meter on the home actually keeps track of on-peak (during the day) and off-peak (late at night) usage separately, and customers pay less for electricity used during off-peak hours. If you set up a timer on your electric water heater and only have it heat up your water at night so you can get up and take a hot shower, throw in a load of clothes, and run your dishwasher in the morning, you won't keep the water heater running all day long. Sounds easy, but don't forget that when you cook dinner, you won't have hot water; you'll need to wash your hands in cold water and it won't be quite as convenient, which is really the only downside. But we've grown pretty accustomed to the comfort and convenience that electricity provides.

What the new Smart Grid will mean is improvements on all of the above items, and much, much more. For you and me, it will mean more economic choice for using electricity off of the power grid in our homes and businesses. For a Smart Grid to become a solid sustainable reality in the future, the distribution systems throughout the country will need to have some pretty significant makeovers, and this includes in your home and every home in the country. American consumers will no longer use or think of electricity in the manner they do now. It's coming, I promise you; I just don't know how quickly it will be accepted. The evolution of it has already started in some areas of the country.

Currently, throughout the majority of the United States and Canada, the population uses electricity with little or no thought. At the end of the month or billing cycle, the electric bill comes in the mail and the recipient of the bill either briefly looks the bill over and pays it or doesn't even look it over and just pays it. In general, little consideration is given to the electric utility bill, because it is fairly constant from one billing cycle to another. We have all been conditioned for decades and decades to realize we pay so many cents per kilowatt used over the course of the billing cycle no matter when we used the electricity. For example, it does not matter if you ran your dryer to dry the towels from 2:00 a.m. to 3:00 a.m., when demand for electricity was extremely low, or if you ran the dryer from 3:00 p.m. to 4:00 p.m. on a 98-degree day when the power grid was stressed to the max and prices for the additional generating stations that do not normally run were operating to support the high customer loads for the extremely hot day. Either way, you, as the consumer, pay the same price per kilowatt used. This is *very* likely to change in the future.

Try to imagine a futuristic world, and allow Eli T. Iceman to help your imagination. Imagine a different meter on your house that communicates to your laptop computer and lets you know every five minutes what the price of electricity is for the next five minutes. Then after 12 five-minute cycles, you get an average kilowatt cost for that hour. You are then billed at the end of the month based all of those one hour time periods. Every hour of every day will have a slightly different cost to you. You will react and think very differently about the electric usage in your home. Technology might provide a digital display you can install in your kitchen that will communicate with the new smart meter on your home and send a price signal reading every five minutes on the display in your kitchen. This digital device might have some settings on it to alert you when the price gets to a certain point to alarm you with some type of audible noise. You hear the noise and decide to turn the air-conditioning off for an hour or so. Or maybe you are cooking a pot roast in the oven and are using the dryer at the same time to dry the sheets from the bed. You decide the price is high and shut both of the appliances off until the prices on your digital meter reduce to a level you are more accustomed to seeing. Maybe the alarm will be set up to text you a message to your cell phone or send you an email if it reaches a certain level.

Just maybe, in the future, the technology gets even smarter than that. Do you recall the big push through our society to buy Energy Star appliances? These are appliances that are designed to use less electricity than their older counterparts. Well, imagine a new initiative being developed to buy Smart-Grid compatible appliances. The possibility exists that the next dryer you buy in the year 2020, for argument's sake, will have technology built in to it. You get the dryer delivered, it's in the laundry room, and you're reading the owner's manual about how to program the highest price you are willing to pay for electricity in a five-minute cycle to dry the clothes. Let's just say you type in 30 cents per kilowatt into the program. Then every time you dry clothes, your new dryer will communicate with your new smart meter every five minutes to find out the price of electricity. If it is more than 30 cents, your dryer automatically shuts off and restarts when the price falls below the 30 cents. Or possibly you go to the hardware store to buy a $10 Smart Grid electrical adapter. You take it home and plug your swimming pool filter into it and set the adapter limit at 20 cents. When the five-minute price gets to 20 cents, the filter shuts off. This all sounds like something out of the old *The Jetsons cartoon show* for any of you that remember that show, but it has got real possibilities to happen. If Neil Armstrong could walk on the moon back in 1969, do you really think the above visions are unthinkable in today's world filled with new technology?

What will all of these changes mean for power grid operations and the industry as a whole? One overall result will be that it will level out the load pattern throughout the day considerably. Why is this important? Simply stated, if the load pattern is relatively flat, the power can be more easily provided with the most efficient resources available for the entire 24-hour day. If people using electricity decide it is too expensive and stop using it while it is expensive, the load reduces and flattens the high points in the load curve.

Another significant contributor to making a smart grid a reality will be businesses. Many industry experts will tell you that huge-scale energy-storage projects are not a cost-effective answer. The better solution to using electricity more efficiently is small-scale projects scattered throughout the electric-distribution system. For example, if battery technology keeps advancing as fast as it has over the past few years, it may very well be possible that a business could store electricity into a battery system at its office during the night when prices are low, then use this stored supply during the day for its computers, lighting and machinery, then recharge the batteries again that night, night after night, week after week. What this will mean for the industry's load pattern is the valleys of the load curve will be partially filled by the electricity charging the batteries for all of these businesses, and top peaks will be lower because the charged batteries are providing what electricity used to. Illustration III-3 shows what the load curve is anticipated to look like when the smart grid is fully integrated.

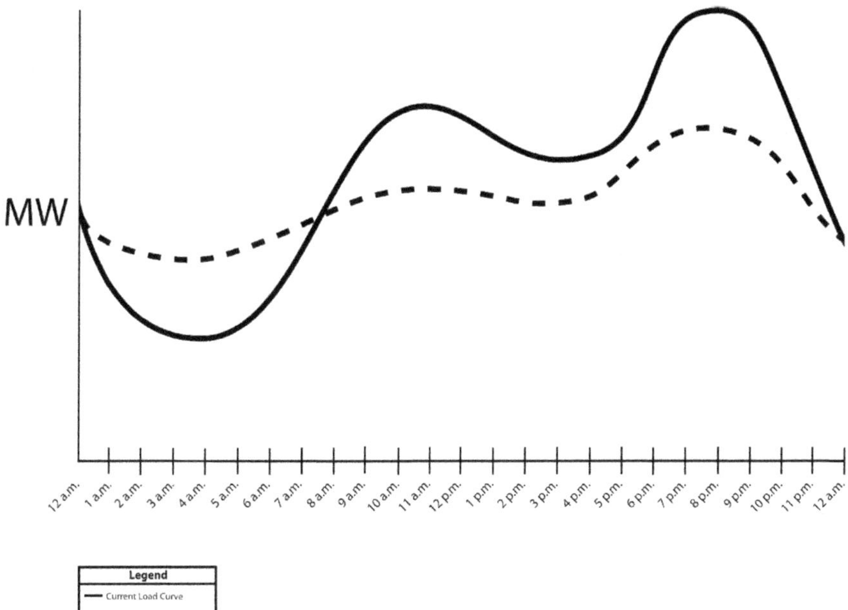

Future Load Curve with Smart Grid vs Current Load Curve

The ultimate goal of most Smart-Grid programs seems to be to have every customer have the ability to have a "smart home" or "smart business" that could be programmed to save the end user money. This is a great concept, but we're still a ways off from it happening nationwide. The industry itself needs to continue to develop to the point that the markets are retail markets instead of the wholesale markets that exist today. As far as the industry's organizations are concerned, this isn't really an insurmountable problem, as it will mean just a few changes for the ISOs and RTOs. It's really more of a challenge for the end users. Today's markets are set up to put a lot of the burden to correctly forecast how much electricity the local utility company will need right back on the individual companies that are members of these large ISOs or RTOs.

Today's electricity markets have financial repercussions for being wrong on the forecast. They have additional charges assessed against companies that use more electricity than they forecasted they would need. They have additional charges if the generator doesn't produce the power the generator was expected to. They have charges if the generator doesn't follow the five-minute instructions it receives. As with many things in life, there are penalties for being wrong, or not following instructions, and the same holds true in the electric markets. This risk will be the same if retail customers begin to get involved in the bidding and "guessing" game of electricity markets as the Smart Grid evolves throughout the nation, and the penalties for being wrong will be shared with the wholesale energy companies.

Emerging Technologies

Similar to the Smart-Grid technology emerging, other technologies are emerging as well, but before we get into them, there are a few points that should be made, which we discuss below.

Oil-Fired Generation—Normally, the most expensive generators on a system run on crude oil that is pretty much about the consistency of tar, so the oil needs to be heated up a lot, then atomized to burn. These types of generators are usually only run on extremely hot or extremely cold days when the overall demand is very high. They are probably some of the least environmentally friendly generators on the grid, as far as emissions, but they only run during extreme conditions and only if absolutely necessary to prevent from shutting off the lights somewhere.

Diesel Generators—These are more efficient than oil generators so are slightly less costly to run but again are relatively *dirty* so normally are selected to run prior to oil, but after everything else.

Simple-Cycle Gas-Fired Combustion Turbines—Combustion turbines (CTs) that run on natural gas are basically just big jet engines that turn generators. The heat generated by the engine itself is wasted, so CTs are the next expensive type of unit normally seen but are not efficient at any load level when they are running. Therefore, the cost difference between the minimum load level and the maximum load level only indicates the cost of burning additional gas.

Combined-Cycle Gas Generator—This type of unit takes one or more simple-cycle gas-fired CTs above and, in addition to the electrical output from the CTs, captures the heat from the exhaust of the jet engines and makes steam with it. This unit then uses the steam to turn a separate steam generator. This greatly improves the efficiency of the overall plant combination and is very clean, relatively speaking, to the others. It is usually the next lowest in the cost stack.

Coal-Fired Generation—Usually the next lowest in the overall cost stack, depending on the technology used at the plant, is coal-fired generation. If the coal-fired unit has not been updated, it can be environmentally unfriendly as well, but most of these units have new and improved upgraded scrubber technologies at their smoke stacks, making them much cleaner in recent years. New government regulations will be forcing the plants with older technology to either install expensive scrubber technology or retire.

Nuclear Generation—Although it is very controversial, nuclear generation is about the cheapest conventional generation in the country to run. It is also normally run at full output all day, every day, to take full advantage of the heat produced in the reactor. Nuclear generation is therefore called base-loaded generation, because as the load across the interconnection goes up and down throughout the day, other types of generators need to increase and decrease their output to make up for the load fluctuations, but nuclear generation stays at full output throughout the day.

Hydro Generation—Hydro generation is one of the favorite types of generation for power grid operators. It is very easy to change the output of the units and is about the most stable, meaning it can take swings on the power system without tripping off. The only hang-up for hydro generation is that a change in water elevation is needed to turn the generators, so either a dam needs to be built and an area flooded to hold the water back and then allow the change in elevation of the bodies of water to spin the generators, or a natural change of elevation has to be taken advantage of. Niagara Falls, on the border between New York and Ontario, is a perfect example of a natural change of elevation, and the New York Power Authority and Ontario have taken full advantage of it for many years. There are hydro generation stations built on both sides of the river that produce enough electricity to power several cities. Manitoba, Canada, has built dams and flooded huge tracts of land in northern Canada and also produces enough electricity to power the entire province and have excess to sell to the US states south of it. Hydro generation is very low cost. There is no cost for the water to turn the generator, only a cost to build and maintain the dams or power plants themselves.

Intermittent Generation—Intermittent generation is just another name for most of the rest of the "green power" resources. Whether wind or solar or any of the other types of electrical power producers, these can only generate when nature allows. A wind generator needs the wind to be blowing hard enough to turn. It also can't turn during extreme wind conditions due to the potential for the generator spinning out of control. Solar generation needs the sun to be shining to allow electricity to be produced. Again, as with hydro generation, there is no cost for the wind or the sunshine, so these resources are also very low-cost resources.

So what's in the future? More wind generation. Yeah, I know, big stretch there, huh? Wind generation is growing across the country faster than any other type of resource. If wind was looked at without considering anything else, it would be perfect. It is absolutely clean, and there are still a lot of locations to put up additional wind farms. It will probably be one of the major suppliers of electricity in the future, but it is not without some operational issues. Until the storage technology advances to the point that we can have more storage than we have wind generation, the cleanliness of wind is *slightly* overrated. Power grid operators get to watch the load pattern of society every single day. They also get to watch the wind-generation pattern every day. More often than not, the wind blows harder in the middle of the night (when the load is lowest) and then begins to reduce as the sun rises in the morning (which is at the same time the load is picking up). Don't let this statement make you think that power grid operators are anti-wind generation, though. Many days, the wind happens, by coincidence, to follow the societal load pattern exactly and may prevent the starting of a generator or even multiple generators for an entire day. Those days are great for everyone, including the environment.

In addition, there are other times the wind is blowing in the middle of the night that causes occasional situations where coal generators will need to reduce their output to allow the "free" wind power to generate. Wind generators often do offer their generation for $0 (and just accept whatever the LMP is), so of course the software the power grid operators use will accept it and reduce the more expensive generators, even though normally the coal generators are some of the cheapest on the system. The wind power gets paid whatever the LMP for its location happens to be set at through the SCUC or SCED software you read about earlier, so everyone's happy, right? Well, sort of.

The coal generator had probably already sold its electricity into the DAM. Let's say it is scheduled to sell 300 MW for the hour from 2:00 a.m. to 3:00 a.m. for $30/MW, then the wind blows a little harder than it was forecasted, so there is 100 MW more wind than anticipated. The coal plant reduces 100 MW, and in the RTM, it buys the 100 MW at $20 to replace the power it sold in the DAM but is now not producing. So the coal plant wins, too; it sold the power in the DAM at $30 and bought it back in RT at $20, so it gets $1,000 profit *and* burns less coal as well. Here's the problem: If, in the morning, as the load starts to increase, and if the wind starts to decrease, as it does many times, the same coal generators will need to pick up for both the load increasing and the wind generation reducing. Remember earlier in the book the discussion about the ability of generators to move up or down at a certain ramp rate? This occasionally causes major operational issues because it actually gets to a point where power grid operators are calling for, and using, all available ramp capability just to keep up with the morning load increasing.

Now, don't forget these areas are required to make sure load and generation are equally balanced, so occasionally, if the wind drops off faster than was forecasted, the power grid operators will likely have to start some offline quick-starting, gas-fired CTs to maintain that generation-to-load balance. Now the market is running a few pretty expensive resources to make up for the drop off of a "free" green resource. This scenario happens more than you would think, and great minds are working together to resolve the problems. This is not a complaint or an opinion; it is simply a fact of what is going on as your alarm clock goes off in the morning and you flip on the bedroom TV and the bathroom light and flush the toilet. There is ***no*** doubt the good about wind generation outweighs the bad, but this is one of the minor downsides.

One answer to resolve this issue is somewhat addressed in the SmartGrid section but will be reinforced here: If electrical storage at the retail level develops and grows, many experts feel this is a possible answer to the above and a lot of other issues. There are a lot of existing technologies for storing electricity, including batteries, electrical flywheels and the power grid operators' personal favorite, pumped-storage hydro. There are also currently negotiations to do some other pretty incredible things.

Batteries–There are currently existing batteries that can store enough electricity to power up to 5,000 homes for as long as 15 or even 30 minutes. There are projects under development that are being advertised as being able to hold as much as 1000 MW and sustain that level for as long as three to four hours. But for the future, my colleagues and I agree with the industry experts and visionaries that batteries will work best and be most cost effective at the retail level. These resources can charge up at night when electricity usage and cost is lowest, then supply the electricity back to the grid when usage and cost for electricity are highest, flattening the traditional load curve.

Flywheels—These are great resources for short-term supply but are not practical for long-term storage. A motor is attached to a large flywheel, and some are rated for up to 1 MW and will consume electricity when the regulation signal from a power grid operator tells them to regulate down (ACE is high), then supply electricity to the grid when told to regulate up (ACE is low). Of course if they are told to regulate up for an extended period, more than five minutes or so, they start to slow down and will need to re-engage the motors eventually to get the flywheels back up to speed. These resources usually net about 0 MW over any extended time period as they flip from providing, then back to consuming, electricity reasonably equally. This technology is not a major source to the power grid currently but does have the potential to grow with time.

Pumped-Storage Hydro—There are generating plants across the country that take advantage of hilly terrain alongside bodies of water to store electricity in a different way. First, a huge ditch is dug up the side of the hill, and some pipe is placed in it that runs from the body of water at the bottom all the way to the top of the hill. Second, a huge pond area is created at the top of the hill—more like a pretty good-sized lake, actually, but it is called a pond or reservoir. Third, a hydro power plant is built at the bottom of the pipe. At night, when electricity usage and prices are low, the plant spins the generators backward and uses them as motors to pump water up the hill to the reservoir. When electrical usage and prices are high, they allow the water to flow back down and spin the generators in the reverse direction to create electricity. It uses slightly more electricity to pump the water uphill than the generator can actually produce, so there are losses associated with this technology, but as long as the price difference between pumping and generating is sufficient to make up for the losses, it is still profitable. This type of technology can support over 1000 MW an hour for 8–10 hours a day. It has extremely fast ramp rates as well, which means it is very high on the power grid operators' list of great generation resources.

There are also currently plans to expand the idea of pumped storage. Some provinces in Canada have huge areas along existing watersheds. Did you know that most of the rivers in Canada flow north? Well, they do. Many of the rivers that start just north of the border flow northward and end up dumping into the Hudson Bay or the Beaufort Sea. But anyway, at night, when wind generation is blowing and sending the price for electricity to low levels, Canadian companies can buy electricity from US companies with excess electricity to supply the customers in Canada. While they are doing this, they reduce generation at their hydro stations and let the water start to back up higher behind the walls of their dams. Then during the day when electrical prices are higher, they can sell more power back to the US companies. This type of system allows both countries to meet higher overall wind usage standards and save money by not needing to run as much expensive generation over the high customer demand periods in the United States. Why not just do the same thing in the United States? Great question;

thanks for asking. The existing river systems in the United States are already being utilized pretty much to the full extent they can be. The watersheds in Canada are measured in millions of acres. If enough water is held back at just one dam to raise the level even one foot, there is enough water to create thousands of MWs for long periods of time. The water that starts flowing north due to rainfall in Winnipeg, Manitoba, takes over a week to get to the Hudson Bay and will flow through five or more dams and generate electricity at every one before it reaches the bay. But even these great plans aren't without issues. Every decade or so, there are drought conditions in which sufficient rain fails to fall to even cover the needs of the Canadian companies, so under these types of conditions, the power grid still needs to have enough conventional generation to meet the overall demand of the system.

The Environment

I am going to admit, I was nearly finished writing this book and just days away from giving it to the editor when I was flipping through the channels on the TV and came across an educational show about our earth. It was a documentary-type program with a scientist narrator named Richard B. Alley. I watched the show with enough interest to go and get the book that the show was partially based on. No disrespect to Richard B. Alley, but I am about as far from being a scientist as I am from the planet Mars. I am an everyday guy, about as common of a person as you will find, and quickly realized the book was over my head, and I wished I could watch the TV show again, as I had missed most of the beginning. Well, even though getting the book proved to be a mistake for me individually, one positive thing came of it: I realized how little I really know about carbon dioxide (CO_2) emissions and the effect they have on global warming.

If I consider myself an average everyday guy, then I have to assume many in my intended reading audience are also reluctant to admit their ignorance about the atmospheric physiochemical issue facing our world; therefore, I did some research so I could share some brief conclusions. I think the proof I did the research is in the fact that I just used the word "physiochemical"! The next few paragraphs are way—*and I mean way*—outside my comfort zone to explain, but I feel I owe it to the reader to try. The electric industry is a contributor to CO_2 emissions, and if I am writing a book about the electric power grid, it seems reasonable I should put a few paragraphs in here about it. I am simply going to point out the major items I learned in my research, in very simple terms. For the record, I should mention that methane (CH_4) apparently causes similar effects as CO_2, but I am going to dodge the methane issue, since it is not power-grid related.

The What, Where, and How of CO_2

Carbon dioxide makes up less than one-tenth of one percent of the atmosphere. CO_2 is produced through breathing by humans and animals. That's not the problem, as natural systems can handle some variations (even increases) in CO_2. The only thing I am aware of that absorbs and uses CO_2 is Earth's vegetation, and it only does this in daylight hours during the plants' photosynthesis cycle, which produces oxygen. Many people don't recognize that plants actually use oxygen to survive, just as animals do, and both release CO_2 in the process. So, during the hours with no sunlight, the photosynthesis process is not occurring and the plants release (respire) small amounts of CO_2 back into the air.

The primary reason for the current concern is that CO_2 is also created in by burning coal. China and the United States are at the top of the global list for CO_2 emissions. In my research, I came across several statistics; these two countries were always on the top of the lists. Sometimes which one listed first was reversed, but everything I read agrees they both belong at the top of the list.

Why is CO_2 Harmful to Earth?

As I am writing this, I can just sense scientists and the coal industry representatives getting in the catlike pounce position to jump all over me and tear me to shreds. I freely admit that I am already over my head on the subject, but here it goes anyway. This is my conclusion from what I read, in the common man's words, not science's words. CO_2 concentrations are measured in parts per million (ppm). Studies and articles I have reviewed in my research consistently state that CO_2 concentrations have risen from 280 ppm to 390 ppm in the atmosphere since about the year 1800, which, coincidentally, was the beginning of the

Industrial Revolution. For you and me, the ppm value does not sound like a lot until you put it in perspective. Earth is estimated to be more than four billion years old, and in just the last 200 years, we have seen the amount of CO_2 in the atmosphere increase by about 40%. So, scientists evaluate what will happen 100 years from now if CO_2 concentrations keep going up at this rate, or 200 years from now, or 500 years from now.

What Happens if We Don't Fix It?

Well, to answer this, we have to know what the CO_2 is doing now. Many argue that CO_2 is causing Earth's atmosphere to warm, which gives us the term "global warming." The term "climate change" has been adopted because it is hard to predict how an overall warming of the planet will affect a specific location. This global warming is argued to be caused by CO_2 molecules that rise into the atmosphere and are spread around by the wind and end up all over the atmosphere, including way up high in the sky. The sun continues to warm Earth, but natural H_2O and CO_2 molecules—and for the last 200 years, the man-made CO_2 molecules—block some of the heat that is trying to rise and move away from the planet. When the heat is blocked, it is redirected to the surface and becomes part of a process called the greenhouse effect. The more CO_2 molecules we put in the atmosphere, the more global warming there will be. What will this do? Forget about it! Absolutely no chance I am going to try to explain the numerous theories that are available to this question.

What Is Being Done to Fix It?

Politicians, world leaders, and industry representatives are all working *somewhat* together to plan for the future. Today, about 50% of the United States' electricity and 70% of China's electricity comes from burning coal. China develops something called a five-year plan, and its 12th five-year plan was released in 2012, stating the country's intentions to reduce its dependence on coal to a lower rate near 63% and to increase other sources with less or no CO_2 effect. This may not sound like a big change to you, but it does amount to a lot less coal being burned to support electricity for the 1.3 billion people in China. This is not big enough in the minds of many, but it is definitely a change in the right direction. The United States continues to build and utilize wind generation, which reduces coal generation as well. But the one thing I learned from the Richard B. Alley TV show was that the United States occasionally struggles with energy policy, as policy in the USA is largely dependent on the elected administration at the time, and often changes with each administration. The CO_2 and global-warming issue is a problem that will take many decades to resolve. It can't be resolved in one elected term, or even two. Policy can be made in that time, but the implementation and effects of the policy will take much longer to measure results. If the next administration changes energy policies, we risk starting over.

What Does It Cost to Run Some of My Appliances?

In the below calculations, I use 10 cents per KW, which is a fairly reasonable value to use, but you can look where you live on the map below, to get a more accurate estimate of the below calculations.

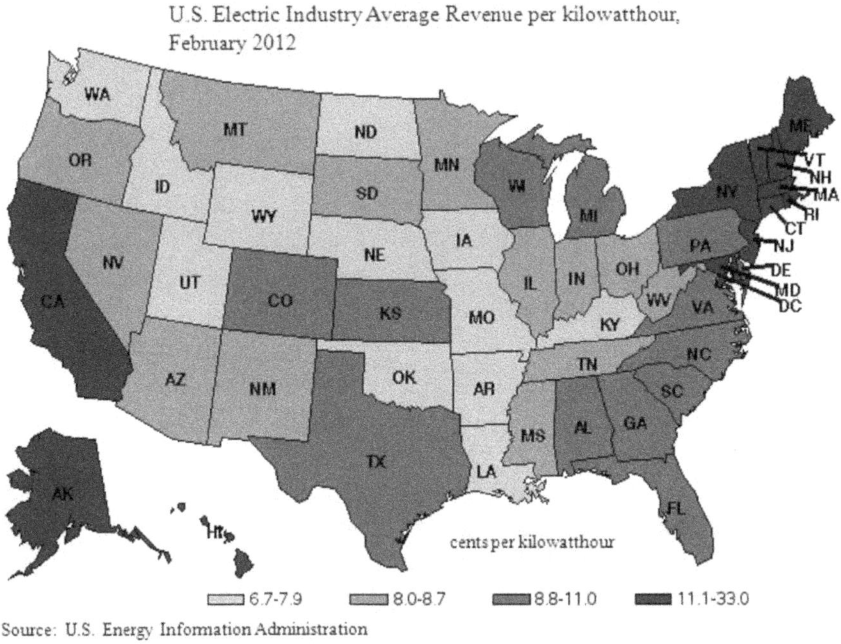

Source: US Energy Information Administration Illustration III-4

Question: How do I mathematically convert amps to watts or watts to amps? I want to do this so I know how much it costs me to use this or that in my house.

Answer: Well, that is a question that begins to scratch the surface of the electrical engineering world. It is a point where young engineers hear a question like this and actually start to get excited. As for a large majority of the general public, we are interested in a different question: How do I convert electric usage to my MONEY? The following examples provide the answer:

Question: My TV says it uses three amps when it is on, so how much does that cost me?

Answer: Assuming your TV is plugged in to a standard 120 volt outlet like millions of other TVs, you take the voltage level and multiply it by the amps: 120 x 3 = 360 watts. So, if your utility company's kW rate is 10 cents, you spent about four cents watching a rerun of *Law and Order*.

Question: I do more laundry than anybody I know. How much does it cost to me to dry a load of towels?

Answer: The main reason this was included was I wanted to make sure you understood there are normally two levels of electricity in your home: 240 voltage and 120 voltage. Dryers normally use 240. So, voltage level multiplied by amps (and the owner's manual of this dryer says the dryer uses eight amps on 240 volts) is 240 x 8 = 1920 watts. That's 1.92 kWh times 10 cents, which means it costs you about 20 cents for a load of towels. But remember: Your bill is calculated on hourly usage; so were the towels in there for an hour, or a little more than an hour, or a little less? Make the adjustments as needed.

Question: I want to buy an electric car and would like to know the cost difference between electric and gas per mile.

Answer: Well, this is an extremely difficult question to answer with a high percentage of accuracy because of all of the variations of gas trucks, cars, hybrids, SUVs and electric cars, not to mention the cost of gas at the time. But I will give it a shot.

I know someone with a small electric car–100% electric charge, no gas backup. He says the car has an advertised range of about 100 miles for a full charge. It takes him about eight hours to charge the car on a 240 volt charger from an empty charge to a full charge. The car draws about 10 amps per hour when charging, which means 10 amps x 240 volts = 2400 watts, which is 2.4 kW. At 10 cents per kilowatt, it costs about 24 cents per hour to charge the car, so eight hours of charging times 24 cents is $1.92 for him to go 100 miles on an electric charge.

Therefore, if you use $4.00 a gallon for gas in a car that goes 20 miles on a gallon of gas, the cost to go 100 miles is about $20.00. In summary, for this scenario:

> 100 miles in an electric car costs $ 1.92
> 100 miles in a 20 mpg vehicle costs $20.00

Things That May Be Discussed Further in the Next Eli T. Iceman Book

There are several things not discussed in this book, or at least not discussed in great detail. To be brutally honest with you, you are light years away from sitting in a power grid operator's chair as was simulated with you as an operator for South Illumini, LLC. The industry is so massive and so technically complicated that I did not even scratch the surface of what was relayed to you through this book, but you should have a basic understanding of how it is operated by now. Several critical areas were intentionally left out because of the added complexity. I hope to possibly include an explanation for some of them in the next book. Some of these subjects are direct current, MVARs, transmission-loading relief, and demand-response programs.

Direct Current (DC) ties between interconnections. Briefly discussed in the Japan section and the Quebec section, DC transmission is a method of tying two interconnections together with specialized equipment. This equipment is not readily available to buy on the shelf of an electrical supply store. This equipment takes years to build and install and is very expensive.

Mega VARs (MVARs) is a subject mentioned briefly in the substations section, but the explanation did not do justice to its importance on the grid. If an analogy could be used, it would be that of pressure on a water system to MVARs on an electric system. Increasing MVARs on an electric system will increase the voltage and more easily push the MWs to where they are needed. I felt the technicality of this subject would bore the reading audience to death, so it was intentionally left out until I can come up with a "fun" way to explain it. To be honest, I spent a lot of time thinking how I could explain it and am not sure I will get there any time soon.

Transmission-Loading Relief (TLR) is a software program designed to remove contracted MWs from the grid because they are causing a potential transmission constraint. This is another tool that power grid operators occasionally use to help them resolve transmission loading issues on the power grid. It is a process nearly every energy trader is aware of because it is one thing that can completely screw up a great deal they organized. TLR is a complicated software package that attempts to fairly curtail contracted schedules from one entity to another when unforeseen problems develop.

Demand-Response Programs are a part of the industry that is rapidly growing. It is a means to provide the grid with help when necessary; these programs are rapidly changing and evolving. I am actually considering focusing the core of the next book in demand-response area in combination with Smart-Grid technology. What the demand-response programs do in a nutshell is this: When a load is anticipated to exceed the available generation resources, the power grid operators can use these demand-response programs to reduce the demand. The programs basically give industrial customers either a slightly lower electric rate or a financial refund in exchange for letting power grid operators terminate their service when necessary. The best phrase to describe this type of prearranged customer load is "optional load." The industry refers to it as non-firm load.

How Do I Become a Power Grid Operator?

Well, if you got to this point reading the book, I would have to say you have a good start. If you enjoyed the book and believe being a power grid operator is a career that might interest you, there is some suggested reading below for a little further explanation of many subjects that were brought up. But before you go and buy a book because you think you want to be a power grid operator, you should ask yourself three questions:

1) Can I work a rotating shift, which means working nights, weekends and holidays?
2) Can I handle after-the-fact criticism for mistakes or decisions I have made?
3) Can I sit in the same room for 8 or 12 hours?

Those are three things that every power grid operator has in common. Now, to be fair, the after-the-fact criticism is handled better by some individuals than it is by others, but you must be able to handle it. The organization and the industry in general want all operators to learn from each other's mistakes, so *when* (not if) you make a mistake, you need to be able to share your experience so your colleagues can learn from it. If you answered yes to all three questions, then keep reading this section.

There is no possible way you can read one of the books below and completely understand what is being explained. These books are in no particular order; these are just books I have personally read, and there are many others at Amazon.com.

- *Electric Power System Basics* by Steven W. Blume
- *Understanding Electric Power Systems* by Delea and Casazza
- *Understanding Today's Electricity Business* by Shively and Ferrare
- *Electric Power Systems* by Alexandra Von Meier

Now, prior to reading one of the books, I strongly suggest that you sign up on LinkedIn social media site and join a couple of groups, such as NERC Certified System Operators and/or Electric Utilities Professionals. This will give you a tremendous resource to use when you have questions about subjects in the book. You can ask a question to these groups on LinkedIn, and people will be more than happy to help you answer it.

So let me ask you, at this point, how much do you have invested financially? Not much. Just a couple of books, and this is a good trial to determine if you believe it is a career for you. If so, then the next steps begin to start costing a little money. These next steps will be accomplished in an effort to obtain a NERC certification. Like I said, it is going to cost a few bucks, none of which I get, so this is good advice; you can trust me on that. I suggest you contact Operator Education Services–North America. You can look them up on the internet at www.oesna.com and inquire as to what type of training they might be able to offer for your situation. The same holds true for another company offering similar services. This company is System Operations Success, which can be found at www.sosintl.com.

Following that, I believe you will need at least one more preparation course before signing up to take the NERC certification exam. Contact Operations-Training-Solutions at www.o-t-s.com. It's a one-man show with an exceptional reputation for excellence. Sign up for one of his certification exam preparation training classes, as this will help you prepare for the exam. I have never met this man but know of his reputation. He is very well thought of by industry representatives throughout all of the interconnections.

After these courses, take some time and discuss with the instructors if they believe you have learned enough of the skills and knowledge to pass the exam. If they say yes, then sign up to take the exam, but be prepared, as there is approximately a $700 charge to do so. If you fail the exam, you get nothing except an opportunity to learn what you need to work on in order to pass the next time, but each attempt will be another $700. Once you get your certification, start applying to some of the hundreds of electric companies and/or wholesale power marketing companies across North America. You can find a list of these at www.nerc.com. Search for their registered membership list to find all of the electric industry entities across North America. For another list of companies possibly interested in a candidate who has a NERC certification, look at http://www.puc.state.tx.us/industry/electric/directories. These two lists may have some overlap, but combined, they list hundreds of companies with good possibilities.

My last suggestion is to contact a recruiter (also known as headhunter) to help you in your search. This service is free to you, and headhunters have a financial incentive to find you a job, as they get paid by the company that hires you, so again, this is free to you. Now, this does not mean you call them and wait. You are still going to have to send out your resume and cover letters, but I believe if you have a good attitude, your NERC certification and the industry training classes you took in preparation for the exam all listed on your resume, then industry representatives will notice your tenacity to break into the industry on your own. The industry, as you have read, is experiencing and will continue to experience an enormous flood of retirements, so there will be openings. There are many recruiters for the industry, but two recruiters who have enjoyed success and specialize in the Operations areas for the Energy Industries are MyCoff, Frye and Prouse LLC,[26] and In Search of Energy Recruiting.[27] It is worth the small investment of time to contact them.

Consider what a four-year college education can cost you. Conservative estimates range from $65,000 to infinity, to yield a good paying job. Your investment into the above approach will be a fraction of that, and, as the Department of Labor states, utility jobs are fairly high-paying.

Good luck to you! Visit www.EliTIceman.com for more fun products.

End Notes

1. Federal Power Commission, Report to the President by the Federal Power Commission on the Power Failure in the Northeastern United States and the Province of Ontario on November 9–10, 1965 (Washington, DC: US Government Printing Office, December 9, 1965).

2. Arizona-Southern California Outages on September 8, 2011, Causes and Recommendations (May 1, 2012), www.nerc.com/page.php?cid=5%7C407

3. Anderson and Geckil (Anderson Consulting), "Northeast Blackout Likely to Reduce US Earnings by $6.4 Billion" (August 19, 2003).

4. Saha and Moody (ICF Consulting), "The Economic Cost of the Blackout" (2003), www.solarstorms.org/ICFBlackout2003pdf.

5. ELCON, "The Economic Impacts of the August 2003 Blackout" (February 9, 2004), www.elcon.org/documents/economicimpactsofaugust2003blackout.pdf

6. United States Department of Labor, Bureau of Labor Statistics, Occupational Outlook Handbook, (February 2012) www.bls.gov/ooh

7. Lauby, "Aging Workforce Webinar" (July 8, 2008) www.nerc.com/fileUploads/file/training/aging-workforce-webinar.pdf

8. NERC Staff, "Long Term Reliability Assessment Emerging issues", http://www.nerc.com/docs/pc/ras/2009_Ras/input_on_LTRA_emerging_issues-compiled_issues2.doc%20-%202009-03-31

9. AEP Staff, Encroachments on Transmission Rights of Way, (Date: Unknown).

10. Casazza and Delea, Understanding Electric Power Systems: An Overview of the Technology, the Marketplace, and Government Regulations, (New Jersey: John Wiley & Sons, 2003).

11. The website for this company is http://www.utilitycomplianceservices.ca/

12. The website for this company is http://www.o-t-s.com/

13. NERC Reliability Standards, http://www.nerc.com/page.php?cid=2|20

14. NERC Program: System Operator Certification, http://www.nerc.com/page.php?cid=6|84

15. For more information, see FERC Order 888/889, www.FERC.gov

16. "The Franz Edelman Award recognizes outstanding examples of innovative operations research that improves organizations—and often changes people's lives. Learning how organizations like yours boost performance through use of operations research—The Science of Better—can help you find new ways to achieve goals in the public and private sector." http://www.scienceofbetter.org/Edelman (2011)

17. NERC Reliability Standards, http://www.nerc.com/page.php?cid=2|20

18. Yong Fu, Mohammad Shahidehpour, Zuyi Li, Security-Constrained Unit Commitment With Constraints (SCUC), IEEE Transactions on Power Systems, Vol. 20 No. 3 (August 2005).

19. Hongyu Wu, Qiaozhu Zhai, Xiaohong Guan, Feng Gao, Hongxing Ye, "Security-Constrained Unit Commitment Based on a Realizable

Energy Delivery Formulation", Mathematical Problems in Engineering, Volume 2012, Article 178193 (November, 2011), 1–22.

20. AK Al-Othman, F.S. Al Fares, and K.M. El-Nagger, "Power System Security Constrained Economic Dispatch Using Real Coded Quantum Inspired Evolution Algorithm," World Academy of Science, Engineering and Technology (2007).

21. The Joint Board on Economic Dispatch for the Northeast Region, "Study and Recommendations Regarding Security Constrained Economic Dispatch (SCED) in the Northeast," Docket No. AD05-13-000 (May 24, 2006).

22. "**Good Utility Practice**—Any of the practices, methods and acts engaged in or approved by a significant portion of the electric utility industry during the relevant time period, or any of the practices, methods and acts which, in the exercise of reasonable judgment in light of the facts known at the time the decision was made, could have been expected to accomplish the desired result of the lowest reasonable cost consistent with good business practices, reliability, safety and expedition. Good Utility Practice is not intended to be limited to the optimum practice, method or act to the exclusion of all others, but rather to be acceptable practices, methods, or acts generally accepted in the region and consistently adhered to by the Transmission Provider." (FERC Glossary, Excerpted from the How Report, Appendix G), www.tsin.com/gloss.html

23. Automatic Generation Control, http://en.wikipedia.org/wiki/Automatic_generation_control

24. NERC Reliability Standards, Standard BAL-002-1 Disturbance Control Performance, www.nerc.com/files/BAL-002-1.pdf

25. Frank Chu Principal, FJC Associates Inc., https://www.gplus.com/utilities/insight/japan-needs-major-power-grid-investment-after-the-quake-53567

26. The website for this company is www.mfpllc.us/

27. The website for this company is www.insearchofenergy.com

Other Sources Used for General Information

The National Commission on Energy Policy, *Ending the Energy Stalemate: A Bipartisan Strategy to Meet America's Energy Challenges* (December 2004).

Steven Warren Blume, *Electric Power System Basics* (2007).

Denise Warkentin-Glen, *Electric Power Industry* (2006).

Z. Li, M. Shahidepour, H. Yamin, *Market Operations in Electric Power Systems: Forecasting, Scheduling, and Risk Management* (March 2002).

Richard B. Alley, *Earth: The Operators' Manual* (April 2011).

CPSIA information can be obtained
at www.ICGtesting.com
Printed in the USA
LVHW021411181218
600890LV00030B/839/P

9 781457 513701